CRC Series
in
MARINE SCIENCE

Editor-in-Chief
J. Robert Moore
Director, Marine Science Institute
The University of Texas
Austin, Texas

CRC HANDBOOK OF MARINE SCIENCE
Editors
F. G. Walton Smith
Dean Emeritus, School of Marine and Atmospheric Science
University of Miami
President, International Oceanographic Foundation
Miami, Florida
Frederick A. Kalbler
President, Hydrobiological Services, Inc.
Naranja, Florida

CRC HANDBOOK OF GEOPHYSICAL EXPLORATION AT SEA
Editor
Richard A. Geyer
Professor Emeritus
Department of Oceanography
Texas A & M University
College Station, Texas

MAN AND THE MARINE ENVIRONMENT
Editor
Robert A. Ragotzkie
Director, Sea Grant Institute
University of Wisconsin
Madison, Wisconsin

CRC HANDBOOK OF MARINE SCIENCE: COMPOUNDS FROM MARINE ORGANISMS
Authors
Joseph T. Baker
Vreni Murphy
Roche Research Institute of Marine Pharmacology
Sydney, Australia

CRC HANDBOOK OF COASTAL PROCESSES AND EROSION
Editor
Paul D. Komar
School of Oceanography
Oregon State University
Corvallis, Oregon

CRC HANDBOOK OF MARICULTURE
Editor
James P. McVey
Fishery Biologist (Aquaculture Specialist)
National Marine Fisheries Service
Office of International Fisheries
Jakarta, Indonesia

Man and the Marine Environment

Editor

Robert A. Ragotzkie, Ph.D.
Director
Sea Grant Institute
University of Wisconsin—Madison
Madison, Wisconsin

CRC Series in Marine Science

Editor-in-Chief

J. Robert Moore, Ph.D.
Professor
Marine Science Institute
The University of Texas
Austin, Texas

CRC Press
Taylor & Francis Group
Boca Raton London New York

CRC Press is an imprint of the
Taylor & Francis Group, an informa business

First published 1983 by CRC Press
Taylor & Francis Group
6000 Broken Sound Parkway NW, Suite 300
Boca Raton, FL 33487-2742

Reissued 2018 by CRC Press

Library of Congress Cataloging in Publication Data
Main entry under title:

Man and the marine environment.

(CRC series in marine science)
Includes bibliographies and index.
1. Ocean. I. Ragotzkie, Robert A. II. Series.
GC21.M265 1983 910'.09162 83-3777
ISBN 0-8493-5759-4

A Library of Congress record exists under LC control number: 83003777

Publisher's Note
The publisher has gone to great lengths to ensure the quality of this reprint but points out that some imperfections in the original copies may be apparent.

ISBN 13: 978-1-315-89514-7 (hbk)
ISBN 13: 978-1-351-07424-7 (ebk)

Visit the Taylor & Francis Web site at http://www.taylorandfrancis.com and the
CRC Press Web site at http://www.crcpress.com

PREFACE

How early man discovered the sea we shall never know. But it probably was not too different from a young child's first experience on an ocean beach: first wonder, then touch, feel, and taste.

The discovery that the sea was a provider of food certainly followed. We can also guess that the invention of primitive craft and the use of the sea as a means of transport were early developments. From simple gathering of shellfish and nearshore fish man soon ventured further from shore and began to explore this alien environment.

Early human cultures that evolved in the marine environment provide striking examples of the astounding adaptability of man to this environment. Polynesian navigators, though lacking in modern scientific technology, traveled long distances between Pacific islands using highly sophisticated techniques of positioning and direction finding. They combined astronomical methods for location with exquisitely accurate sensing of wave motions for establishing and holding a course. They used sails and the wind for propulsion. Because wind is driven by the differential heating of the earth by the sun, the Polynesians were among the first people to tap solar energy. Their shipbuilding skills were highly developed and were uniquely adapted to provide safe and reliable craft for the long voyages they undertook. Though these people did not have what we consider to be a scientific understanding of the ocean, they more than made up for this lack by their powers of observation and their tradition of passing on their knowledge from generation to generation.

The Innuit or Eskimo developed a somewhat different approach to the sea, adapted to their polar environment. They too learned to navigate but their travels were mostly over the frozen ocean. They made extensive use of sky appearance to find the open water and leads where marine mammals like seals, walruses, and even whales were to be found. There are many words in the Innuit language to describe the sky and what its appearance means in terms of the underlying ice, snow or water. Even today the skills of the Eskimo in pursuing and killing bowhead whales in the open sea from tiny skin-covered kyaks are convincing evidence of man's ability to deal with the marine environment under harsh and primitive conditions.

In Hawaii, aquaculture reached a high stage of development centuries ago in the form of great fish pens built to capture young fish larvae and raise them for food for royal families. Many of the problems which plague modern attempts at aquaculture were eliminated by these pens: capture of fish larvae, natural feeding, water treatment by constant flushing of sea water through the permeable rock walls, exclusion of predators by these same rock walls, and continuous availability of the fish as needed. Besides being an effective and efficient means of providing food, these extensive pens built along the coast were an early example of what today we call coastal management. A portion of the coast was allocated for an exclusive use by a particular group, in this case the royal family, and was managed solely for this purpose.

The Native Americans of the southeast U.S. though not a marine-based culture, traveled to the sea islands each summer. There they harvested and consumed hugh quantities of oysters, clams, and mussels. The large doughnut shaped piles of shell, some 50 m across, which they left, suggest regular campfire sites where their "summer holidays" took place. Mild climate, abundant food, and pleasant surroundings, which still today attract large numbers of recreation-seeking people to these islands, were no doubt a welcome respite from hunting and farming on the mainland.

In examining contemporary man and his relation to the marine environment one is immediately faced with the question of which man and which marine environment? Certainly the view of the sea by the working diver in the cold and dark water several hundred feet below the surface of the North Sea is hardly comparable to that of the surfer riding a wave on a Hawaiian beach. The captain of a merchant ship sees the sea quite differently from the weekend skipper of a day sailer out of Newport. Yet all these people are interacting with some part of the marine environment.

This book then is not a unified treatise on how man interacts with the sea, but rather a collection of accounts of how different parts of present day human society perceive and deal with the marine environment.

Niels Rorholm, an avid sailor as well as a highly respected resource economist, examines the subject of marine recreation. As he points out " . . . the bulk of water-related active recreation was once, and frequently still is, done by others as work or for survival." Sailing, diving, fishing, and even surfing all originated as work activities. Man's need for play and the mystical attraction of the sea are clearly reflected in the large and growing economic value of marine recreation which in the U.S. generated expenditures in 1979 of over $23 billion, nearly as much as the fisheries industry. Marine recreation has taken a place of prominence in the list of human activities which are straining the resources of the sea, especially the seacoast, to support.

Courtland Smith and William Wick introduce us to the anthropological origins of fishing for food. Their sensitive account of how human culture and fishing are intertwined provides an exceptional insight to why attempts to manage fisheries have so often met with failure or a holding action at best. The change from artisanal fishing to well capitalized and technologically advanced fishing fleets is causing us to outrun the sea's capacity to provide. Beyond man's over efficient exploitation of the living resources of the sea the authors see a future where human colonization of the seacoast and the development of offshore communities will become the major threat to fishing as it is known today.

Water transport has changed the face of the world. It has made possible trading and commerce between distant peoples, and it has been the basis for the colonization of north and south American continents as well as the Pacific's islands and the subcontinent of Australia. Harry Brockel, seaman, port director, and scholar, traces the process of the opening of the agricultural and industrial heartland of the U.S. and Canada and shows how its development was totally dependent on and integrated with the development of a Great Lakes shipping industry. He explains how this region of unparalleled productivity was and remains dependent on and integrated with a strong and innovative Great Lakes shipping system. From the first use of propellers to drive steamships and the pioneering of container ships to the present day 1,000 ft bulk carriers used for iron ore, coal, and grain the Great Lakes shipping industry has continued to evolve to meet the needs of the region.

As a land animal man has, until recently, been prevented from occupying the undersea environment. Edward Lanphier, physician and research scientist, tells us about the physiological constraints which even today make the underwater environment a hostile one for man. Technological adaptations have overcome many of these constraints, but the barriers of depth and high pressure and the strange effects of various gas mixtures on divers continue to be the subject of intense and sometimes risky research.

Despite elaborate and sophisticated technological advances, work undersea is still best accomplished by a man on the site, be the task the recovery of a lost object in 30 feet of water or the final placement and connection of the "Christmas tree" on a completed oil well 1,000 feet beneath the surface of the sea.

The vast array of techniques and systems used by commercial divers are graphically described by William Hamilton, an ocean-science consultant who deals primarily with decompression procedures and related matters of advanced underwater physiology. Modern commercial diving has developed primarily in response to the needs of the petroleum industry. The requirements of the offshore oil production industry are increasing rapidly and the 1,000 foot depths of today will no doubt be exceeded in years to come. Whether humans can safely operate at the ambient pressure of such depth or whether manned submersibles or anthropomorphic diving suits, both maintained at one atmosphere pressure, will replace the free or tethered diver remains to be seen.

With the rapid and apparently limitless progress of modern technology and engineering the ability of man to cope with and control the forces of the marine environment seems to be only a matter of money and persistence. The fallacy of this belief is convincingly demonstrated in Orrin Pilkey's chapter on coastal erosion. As a marine geologist Pilkey carefully explains how the sea and the coast interact to achieve a stable and predictable balance. The reality of this balance has been documented time and time again. Yet despite this knowledge, man and his institutionalized agencies, particularly the U.S. Army Corps of Engineers, persist in trying to control coastal erosion by structures designed to stop or reverse the natural coastal processes. That these attempts to control the forces of the sea have little or no effect on the natural processes and often exacerbate their effects in the bargain suggests that scientific understanding by itself is not enough. The problem here, as in most other marine environmental issues, is man himself. The difficulty of accepting the natural course of events whether it be coastal erosion or the reality of the marine ecosystem continues to confound man and his institutions.

The passage of the Coastal Zone Management Act in 1972 marked the beginning of a major experiment to achieve a revolutionary change in our institutional approach to the management and conservation of the coastal environment. For the first time, the federal government undertook to deal with a major part of the marine environment in a comprehensive way. Allen Miller, Administrator of the Wisconsin Coastal Management Program, traces this experiment from the findings and intent of the U.S. Congress to the gradual agreement of the majority of the coastal states to enter into a partnership with the federal government to manage their coastal zones as a whole. The brief and controversial history of this program exposes the positions of the special interest groups almost all of which see a loss of power or profit in the holistic approach. Whether this particular experiment will survive in its present form is not at all certain at this time. However, the public awakening to the values and complexity of the coast and the broad public support for the more effective use and conservation of the resources of our coasts are real achievements which will not disappear.

The legal and philosophical basis for public management of the coast including such questions as public access, private property, the implications of erosion or accretion of land, and conflicting uses has been dealt with mostly on a case by case basis. With the advent of the Coastal Zone Management Act the frequency of such cases has risen sharply. David Owens, an attorney for the North Carolina Coastal Management Program, has examined this subject in detail. He gives a scholarly yet fascinating account of who owns coastal resources and how governments at the federal, state, and local level are attempting to manage and regulate these resources. He also discusses the constitutional issues raised by these regulations.

In the last chapter the Great Lakes are likened to a microcosm of the world ocean. In this region of concentrated population and intense industrialization overexploitation of marine resources and degradation of the marine environment have occurred much quicker than in the ocean. Human and institutional response has also been faster. The level of public understanding of the issues is probably higher than in the case of ocean resources and real progress has been made in resolving some of them. Much can be learned from the Great Lakes experience, both scientifically and institutionally.

Though this book represents only a sampling of the wide variety of man's interactions with the sea, it is nevertheless broadly representative. It ranges from man's psychological and aesthetic needs, his physiological limits and technological adaptations to his societal relation to the sea in terms of legal, economic and institutional interactions. If there is a common thread which weaves all these things together, it is the inherent limits of man himself in trying to deal with this complex and alien environment. Despite a high level of biological adaptability and sophisticated technological adaptability, man is still constrained by his bodily limits and the imperfections of his devices and structures. The sea's lack of forgiveness of human weaknesses remains a continuing reminder of man's subordinate role.

As a social animal man has developed an extremely diverse and imaginative array of social organizations and institutional arrangements to deal with his environment. Despite this social and institutional flexibility, man has not yet learned to live in harmony with the sea. Primitive cultures probably came closer to this ideal than contemporary man, but owing to their limited population and low level of technology theirs was a simpler and less demanding situation than mankind faces today.

As the potential for man's exploitation of marine resources and control over his environment has increased, the apparently infinite capacity of the sea has become finite. Man's ability to achieve harmony and balance with the marine environment, indeed with his total environment, is being thwarted by his greed and desire for power. Social, economic, and political conflicts are proving to be more difficult to overcome than the challenge of the sea itself.

This collection of essays strives to awaken in the reader an appreciation for the many ways man is involved with the sea and the limitations to his dominance over this environment. It also attempts to show how man, wherever he lives on the earth, is ultimately dependent on and involved with the sea.

Lastly, it is important to note that the authors of this book are all people who are personally involved with the marine environment. They all dive, fish, sail, or work in the sea. Their views of man and the marine environment are based on lifelong observation and study. Their love for the sea stems from their understanding as well as from their human response to this mysterious part of our world.

Robert A. Ragotzkie
Madison, Wisconsin

EDITOR-IN-CHIEF

The Editor-in-Chief of the CRC Marine Science Series, **Professor J. Robert Moore,** received his B.S. degree (honors) from the University of Houston, in geology, his M.A. from Harvard University, also in geology, and his Ph.D. from the University of Wales (Aberystwyth), in geology and oceanography. From 1956 to 1966, he was a Senior Scientist at the Texaco Research Laboratories, Houston, Texas, engaged in research on marine sedimentation and geo-resources. From 1966 to 1977, Dr. Moore was Professor of Geology at the University of Wisconsin, Madison, where he was also Director of the Marine Research Laboratory and the Marine Mining Program. In 1977, he became Director, Institute of Marine Science, University of Alaska, and in 1979, assumed his present affiliation as Professor of Marine Studies, at the University of Texas, at Austin. Professor Moore's principal research interests are in seafloor mineral resources, marine geology, and global oceanographic processes and resources. His research has involved studies in the Bering Sea, Irish Sea, Western Pacific, Gulf of Mexico, Gulf of Alaska, Atlantic shelf and several coastal areas in the United States and abroad. He is a member of the Marine Board of the National Academy, the A.A.P.G. Marine Geology Committee, the N.A.C.O.A. Marine Mining Panel, Organizing Chairman of the CHEMRAWN IV Conference, and member of The Geochemical Society, Soc. of Economic Paleontologists and Mineralogists, Marine Technology Society, Am. Assoc. of Petroleum Geologists, Challenger Society, and a Fellow of the Geological Society of London. He is Editor of the *Marine Mining Journal,* and Exec. Secretary of the International Marine Mining Association. Since 1966, he has been an active consultant to major international corporations, chiefly on matters related to ocean resources, utilization, exploration, and development. He is currently conducting two major marine research programs, one in Palauan waters of the Western Pacific and the other in the Bering Sea off Alaska. He resides in Austin, Texas.

THE EDITOR

Robert A. Ragotzkie, Ph.D., is Director of the Sea Grant Institute and Professor of Meteorology and Environmental Studies, University of Wisconsin, Madison, Wisconsin.

Dr. Ragotzkie earned his Ph.D. in Meteorology and Zoology at the University of Wisconsin-Madison in 1953. He established and was the first Director of the University of Georgia Marine Institute at Sapelo Island, Georgia. His work there was primarily on estuaries and the ecology of salt marshes. After returning to the University of Wisconsin in 1959 as a member of the Department of Meteorology he organized the Marine Studies Center and subsequently initiated the University of Wisconsin Sea Grant Program which he now directs.

Among his numerous professional society memberships Dr. Ragotzkie is an active member and past President of the International Association of Great Lakes Research and is an Associate Editor of the Journal of Great Lakes Research. He is also on the Editorial Board of Water, Air, and Soil Pollution. He has published over 50 scientific articles ranging from physical limnology to the ecology of the loggerhead turtle.

Dr. Ragotzkie's present research interests are in Great Lakes limnology and the heat budgets of lakes. His national activities include past President of the Sea Grant Association, Chairman of the National Council of Sea Grant Directors, and charter member of the Board of Directors of the Marine Division of the National Association of State Universities and Land Grant Colleges.

CONTRIBUTORS

Harry C. Brockel (Deceased)
Lecturer
Center for Great Lakes Studies
University of Wisconsin-Milwaukee
Milwaukee, Wisconsin

Robert W. Hamilton, Ph.D.
Consultant
Decompression and Diving Physiology
Hamilton Research, Ltd.
Tarrytown, New York

The Rev. Edward H. Lanphier, M.D.
Senior Scientist
Department of Preventive Medicine
Assistant Director for Research
The Biotron
University of Wisconsin-Madison
Madison, Wisconsin

Harold M. Mayer, Ph.D.
Professor
Department of Geography
Center for Great Lakes Studies
University of Wisconsin-Milwaukee
Milwaukee, Wisconsin

Allen Miller
Program Manager
Wisconsin Coastal Management Program
Madison, Wisconsin

William J. Neal, Ph.D.
Professor
Department of Geology
Grand Valley State College
Allendale, Michigan

David W. Owens, Esq.
Assistant Director
Office of Coastal Management
North Carolina Department of Natural Resources and Community Development
Raleigh, North Carolina

Orrin H. Pilkey, Ph.D.
Professor
Department of Geology
Duke University
Durham, North Carolina

Robert A. Ragotzkie, Ph.D.
Professor
Department of Meteorology and Institute of Environmental Studies
Director
Sea Grant Institute
University of Wisconsin-Madison
Madison, Wisconsin

Niels Rorholm, Ph.D.
Professor of Resource Economics
Coordinator
Sea Grant College Program
University of Rhode Island
Kingston, Rhode Island

Eric Schenker, Ph.D.
Dean
School of Business Administration
University of Wisconsin-Milwaukee
Milwaukee, Wisconsin

Courtland L. Smith, Ph.D.
Professor
Department of Anthropology
Oregon State University
Corvallis, Oregon

William Q. Wick, M.S.
Director
Sea Grant College Program
Oregon State University
Corvallis, Oregon

To Harry Brockel, sailor, port director, and teacher to us all, this book is respectfully and fondly dedicated.

Harry Brockel with his ship people on the docks of the Port of Milwaukee. (Milwaukee Journal Photo)

TABLE OF CONTENTS

Chapter 1

MARINE RECREATION

Niels Rorholm

TABLE OF CONTENTS

I. INTRODUCTION

Marine recreation represents one of the most important uses of our marine and coastal environments.* This is a natural result of three things: population density in coastal areas, increasing interest in outdoor recreation throughout the country, and the special lure of the sea. What the poet may think of as the lure of the sea, today's recreation planner would consider a combination of resource-directed and image-directed desires.[1]

According to the 1980 Census of Population, 80% of Americans live in coastal or Great Lakes states on 56% of the U.S. land area, meaning higher population density along the coast. Reading the third nationwide Outdoor Recreation Plan one learns that "From all walks of life, 59% of all Americans regard outdoor recreation as very important compared to their other interests. An additional 24% consider (it) to be somewhat important."[2] The denser coastal population has a lot of interest in outdoor recreation.

Categorizing and quantifying the lure of the sea is more complex, but evidence indicates it is there. It is not hard to understand the attraction of escaping the city on a hot summer's day and taking a refreshing ocean swim or going for a sail. But it taxes many people's imaginations to see the charm of lying dressed in a wet suit on a surfboard waiting for a wave in the winter North Atlantic. Similarly, crew members on particularly "brisk" sailboat races from Newport to Bermuda must have to remind themselves occasionally that they are doing it for pleasure, and yet they come back next time. Other evidence of the special attractions held by the marine environment may be found in the considerable sums people are willing to pay to live year-round close to the coast. Similar houses command much higher prices when located on the coast than if they were a block or two from the water. That extra price paid can only be considered a payment for "recreational services received" from the marine environment.

Now, if variety is a basic human need, then much of this makes sense. For, in the many forms of marine recreation, people tend to do things that are quite different from what they do at work. And in the active forms, the body as well as the mind gets a workout. While, in our working environment, decision-making is becoming more and more decentralized, when diving, surfing, water skiing, sailing, or game fishing, the individual is making the decisions, making them fast, and taking the consequences (Figure 1). And seeing the consequences rapidly reflects another important difference from today's common working environment. At work we frequently do not learn whether our specific decisions were the right ones for a long time, if ever.

In the passive forms of marine recreation the sensory stimulus ranges from the euphoria of floating on a barely undulating sea surface which truly suggests the ocean as the cradle of life, to the bracing experience of absorbing the onslaught of sounds and sights that are a coastal storm.

So, because of a general and increasing bent toward outdoor recreation, higher populations in coastal states and the lure of the sea combined with ability to travel, marine recreation is a growing activity which supports a growing commercial sector and places increasing pressure on our limited coastal areas.

But, before getting into that, perhaps one should ask, "What is meant by marine recreation?"

II. MARINE RECREATION AND VARIETY

Broadly speaking, we may think of marine recreation as "activities other than work-for-pay that depend on the marine environment for their recreational values." So, does the eating of a lobster dinner in Ottumwa, Iowa qualify as marine recreation? No. A product

* When used without qualifiers, the term "marine" here refers to the oceans and the Great Lakes.

FIGURE 1. A refreshing sail on a hot summer's day.

of the sea is involved, but not the marine environment itself. What about dinner in a Seattle waterfront restaurant? Yes, chances are the restaurant is selling not only food, but the "waterfront atmosphere" as well.

As is probably the case with most outdoor recreation, the bulk of water-related active recreation was once, and frequently still is, done by others as work or for survival. Sailing, for example, has a long history of association with hard work and necessary transport. The skills of surfing were surely necessary in returning canoes with or without outriggers safely to beaches after work or travel. Diving was (and is) a way of harvesting some marine species such as sponges and abalone. This connection with former lines of work has created a broader dimension to the affected forms of recreation through an increased interest in the history of what are now hobbies. We find groups restoring old vessels, learning to build wooden boats, and increasingly attempting to navigate without modern devices even as (or perhaps because) these devices are becoming increasingly available.

More than any other natural-resource-related spectrum of recreation, that which is based on marine resources has variety. The variety of marine recreation probably goes a long way toward explaining its broad appeal (Figures 2 and 3). One can be physically passive as in sunning, or extremely active as in competitive swimming, surfing, or sailing. Perhaps the most passive of all, placing one's residence by the shore, is also the form of marine recreation for which the "participant" pays most dearly. A group of Sea Grant Marine Recreation Specialists compiled Table 1 showing the kinds of marine recreation they could think of.[3]

It is interesting to note that the recreation experts did not list a coastal location for a year-round residence as recreation. Yet, if the air, or the view, or the closeness of beach or boat did not serve recreating purposes, why would people pay such enormous sums for the privilege of living in that environment?

The variety possible in marine recreation has several effects on participants, some beneficial, some not. First, it makes family participation more likely. For example, in many areas, some family members can fish while others swim. The range of activities also means that there will be some that are suitable for almost any age group, and there are strenuous as well as leisurely pursuits available.

On the other hand, the very benefits bestowed by our broadly usable coastal resources bring their own kind of problems. If a beach is at once usable for several different activities, chances are that conflicts among them will arise that will tend to lower per capita benefits from the beach and could end up lowering total benefits. A volleyball game may impose only mild annoyance on people who are sunbathing close by. Similarly, sand sprayed from running feet is not an unexpected ingredient in sandwiches at the beach. Such conflicts among activities may be minor, but they do sometimes turn intended recreation into tension and annoyance. Problems caused by mixing of uses such as swimming, surfing, and water skiing cause more obvious conflict. This mix carries the potential for injury, and governmental agencies have had to step in and designate areas for permitted uses.

But this potential for conflict, since it is a reflection of the suitability of the resource (beach and water) for several purposes simultaneously, also means that the potential payoff from even a minimum of planning and regulation is great. All that is needed is the separation of the activities either in time or space. In the example above, spatial separation would likely be preferred. But we will return to the matter of resolving conflicts later.

It is likely that more planning and regulation will be a part of marine recreation's future. To most people that is not a happy prospect. After all, we turn to the water to rid ourselves of constraints, be they clothing, crowds, traffic, or rules.

But, the freedom coin has two sides, one is freedom to do as we please within the law, an ideal cherished by many. The reverse is freedom from being imposed upon by others who are doing what *they* please — the "freedom to, freedom from" conundrum. Increasing one kind of freedom will automatically decrease the other. If you want to be free of loud rock music on the beach, then I must be asked (told) to use earphones or play my radio at home. If I am free to play it on the beach, you are not free to have peace and quiet. Finding the balance which will result in maximum use of our scarce recreational space without unduly decreasing recreational values is the delicate task of those who make coastal land and water use policies, and of those who implement the policies. As the intensity of use of marine recreational resources increases, the freedom trade off becomes a zero-sum game. That is to say, if someone wins, someone must lose an equal amount.

III. THE DEMAND FOR MARINE RECREATION

Research data describing marine recreation are not plentiful, nor have they all been

FIGURES 2 and 3. Marine recreation has variety.

FIGURE 3. Photo by Mike Tsukamoto.

Table 1
RECREATIONAL ACTIVITIES IN COASTAL AREAS

On water	In/under water	Beside water
Boating (motor, sail, row)	Swimming	Tourism + sightseeing
Cruising (gunkholing, rafting)	Body surfing	Picnicking/barbecueing
Racing	Skin diving	Clambakes
Water skiing	Scuba diving	Camping
Tow gliding	Underwater photography	Bicycling
Wind surfing	Spearfishing	Jogging + walking
Water cycling	Underwater archeology	Hang gliding
Jet skiing	Recreational mini-submarining	Kite flying
Canoeing	Wading	Bird watching
Sport fishing (trolling, gamefishing, etc.)	Shellfishing	People watching
Partyboat fishing (head + charterboats)	Crabbing	Waterfowl hunting
Waterfowl hunting	Seineing	Running dune buggies
Harbor/bay cruising (sightseeing)	Animal/shell collecting	Volleyball, football, frisbee, soccer + other sports
Surfing		Beachcombing + treasure hunting
Skipping stones		Beach going
Ice boating		Sunbathing
Skating		Sand castle building + burrowing
Ice fishing		Tide pool observation
Whale watching		Fishing (fly + surf casting, etc.)
		Hearing/seeing/smelling the water waves, storm + watching sunsets
		Boat building

Note: Beyond the obvious swimming, boating and fishing, many people and government officials are hard pressed to identify other ways people recreate in coastal areas. This list attempts to identify diverse activities ranging along our shores beside, in, on and under the water. Consider this to be a working document to be used, edited, reorganized, and expanded as needed.

(From Sea Grant Situation Statement and Recommendation on Marine Recreation, Sea Grant Meetings, Newport, R.I., 1977 and Norfolk, Va., 1978.)

collected in comparable ways over the years. But the story they tell about the increasing pressure of people on our nonincreasing length of shoreline is compelling and conforms to observations of increasing crowds at beaches and on waterways.

Studies of outdoor recreation participation were made in 1960, 1965, 1972, and 1977. There are some problems in comparing the data over time, in that samples and definitions have changed. However, it appears that, except for swimming, the 1960 to 1977 comparison gives a reasonably valid view of the changes that have taken place.

Table 2 indicates how participation rates have increased over the 17-year period for important forms of marine recreation.

Considering that the 1977 swimming data did not include pool swimming, it is safe to say that rates of participation in all these water-dependent forms of outdoor recreation increased significantly over the period (Table 2).

The above data discuss the *rate* of participation. We have already shown that the population density is higher in the coastal and Great Lakes states than it is in the inland states. The increased rate and the increased density together suggest a greater-than-average participation in water recreation in coastal states. Confirming this pressure on the marine resources are data that show the proportion of the U.S. population living in coastal counties:

Table 2
PERCENT OF RESPONDENTS PARTICIPATING IN SPECIFIED OUTDOOR RECREATION ACTIVITIES

	1960	1977
Swimming[a]	45	46
Fishing	29	53
Sailing	2	11
Other boating	22	34
Water skiing	6	16

[a] Pool swimming was included in 1960 but not in 1977.

(From *The Assessment*, Third Nationwide Recreation Plans, Department of Interior, Washington, D.C., 1979, 41.)

PERCENTAGE OF U.S. POPULATION LIVING IN COASTAL COUNTIES (SELECTED YEARS 1800-1960)

1800 — 37%
1850 — 25%
1900 — 30%
1950 — 44%
1960 — 45%

(From *Our Nation and the Sea*, Report of the Commission on Marine Science, Engineering and Resources - USGPO, Washington, D.C., January 1969.)

Studying the conterminous U.S. coastal counties, that is to say, excluding Alaska, Hawaii, and the 85 counties around the Great Lakes, Niels West determined that the 286 U.S. conterminous coastal counties contained 31.9% of the population in 1960. This grew to 33.5% by 1974.[4]

Most of our major cities are located in coastal and Great Lakes counties. These population centers result in pockets of immense density of coastal use. Also, a large city generates its own suburbia which, in turn, generates a semi-rural lifestyle further away from the city center. Therefore, a major city influences the population density in a very large area. If it is a coastal city, this spreading effect is likely to move along the shore partly because of its desirability, and partly because the water prevents a 360° dispersion of people from the city. The result, again, is increasing coastal density. West notes: "In some respects, the coastal rural counties provided the real surprise. As a group, these counties grew slightly faster than the coastal urban counties during the 1950's and increased their rates during the 1960's. Of even greater interest, however, has been the ability of these counties to continue to grow during the early 1970's at rates two to three times those of the two urban classes (of counties)."[4]

The inflexibility of our coastline is another cause of crowding. While some remedies such as fishing piers can, in a sense, extend the coastline or at least increase the capacity of a given length of it, and while man-made islands or canals can in fact make it longer, people primarily deal with the problem of crowding by traveling further to reach less crowded shore and by shifting land from private to public use.

There is ample evidence that people have been willing and able to travel further to their favorite recreation. On the east coast, the evidence is on Route 50 between Washington, D.C. and Ocean City, Maryland; in seemingly isolated clusters of pleasure boats in the

Carolinas; in the predominance of Philadelphia ownership of boats moored on Maryland's eastern shore, and on the small air fields on the coastal islands of Georgia, Rhode Island, and Massachusetts. But the fact that people are pushing against income levels to enjoy their favorite marine recreation is perhaps best illustrated by the carload of work-weary New Yorkers arriving on Rhode Island or Cape Cod beaches late Friday night for a weekend of fishing for striped bass. Or by those showing up before dawn at Manasquan, Montauk, or Point Judith for a day of fishing on one of the "head boats". That type of recreational fishing is not associated with the higher incomes of those who own real property on the shore or have boats with overnight accommodations located well away from their residences.

IV. THE SUPPLY OF MARINE RECREATION

The increase in supply of recreation accomplished by shifting land from private to public uses has taken two or, strictly speaking, three forms.

The borderline case is the division of private coastal lands into house lots. It increases coastal access, but it is still private access. The process is responsible for the suburbanization of fairly large coastal areas, notably in California, Florida, and southern New England-New York.

A second way to shift land from private to public use is to build hotels, motels, resorts or marinas on formerly private lands. In a strict sense the lands are still private, but they now serve public purposes. It is worth noting that typically it has not been economically feasible for a marina to bid land away from private uses unless the marina is connected with residential uses such as apartments, hotels, or condominiums, or is involved with boat sales or other businesses.[5]

Another way of putting that is to say that if a marina for 200 boats requires, say, four lots of shore land, then the benefits that four families feel they receive from living on the shore exceeds the profit potential from storing and servicing 200 recreational boats (assuming a perfectly functioning land market). If they did not, then the great need for boat storage spaces would cause a potential marina owner to bid the land away from residential use.

The third and purest way of transferring coastal use to the public is for public bodies to purchase the land and hold it. A good many small beach areas have historically been held by towns or counties for use by residents and the right of access to the shore for fishing and gathering of seaweed or driftwood. Since 1950, much recreational land has passed to nonprofit ownership such as the Conservation Foundation and the Audubon Society. Frequently such nonprofit organizations have functioned as conduits, assisting states or other levels of government in acquiring coastal lands and islands for public benefits, but not always for public use, as in the case of wildlife preserves.

Looking ahead to the 1980's, one would hazard the opinion that the acquisition of lands, whether coastal or inland, for public recreation is going to decline.

From the 1971 "National Shoreline Study" of the U.S. Army Corps of Engineers, we see that 70% of the U.S. shoreline (excluding Alaska) is in private ownership. The Federal Government owns 11% and 9% is owned by state or local governments. And the study shows that very little of the coast is developed. Again, excluding Alaska, where 99% of the shoreline is undeveloped, indications are that U.S. coastal use is as follows:

	Miles	Percent
Undeveloped	22,500	60
Developed for nonrecreational use	6,000	16
Public recreation	3,400	9
Private recreation	5,800	15
Total	37,700	100

FIGURE 4. Marine recreation is putting heavy pressure on associated resources.

Sixty percent of our coasts are in their natural state. This seems a lot to some people and not enough to satisfy others, but compared to the coast dedicated to public recreation, it is a huge resource. Remember also that "undeveloped" frequently means recreational to those who can find their way on foot, by water, air, or even by automobile.

With rising energy costs it may be difficult for many to continue the enjoyment of those remote coastal areas, but therein lies, perhaps, the salvation of our natural coastline.

V. RECREATION AND RESOURCES

As one counts miles and assesses interest, it is important to realize that even as marine recreation varies from a child fishing with a stick, line, hook, and worm, to the family and guests voyaging on a 100-ft yacht, so do the resource requirements vary from what can be found for free to the expenditure of considerable time, organization, and millions of dollars. Length of coastline may have very little to do with the amounts of diverse recreation available to people, so we need to look at resource requirements of different kinds of recreation, and this is very difficult information to obtain.

The point has been made that marine recreation activities in many cases are putting heavy pressure on the associated resources and that there is a great variety in human perceptions of the recreation offered by a given coastal setting (Figure 4).

The point has also been made that partly because of these two facts there could be high pay-off from a minimum of planning. But the information needed for this planning is elusive. Only a few of the economic sectors contributing to marine recreation are accounted for separately in our government's routine collection of statistics. An example is "boat dealers", but even that grouping is questionable for the dealers also sell boats for nonrecreational purposes, and not all recreational craft are sold through boat dealers. Most examples are worse. Most supplies for marine recreation are mixed in with such sectors as transportation services, miscellaneous plastics, etc., and the marine share of the data are either not available at all or not on a state level. This makes it difficult to get a picture of the field for planning purposes, and supply and demand characteristics can be obtained only through fairly small-sample case studies.

Let the matter of beach capacity serve as an example of the importance of data on resource requirements for planning. The Bureau of Outdoor Recreation has developed standards describing the area of beach needed to avoid undue crowding. The figure has been set at 75 square feet of beach per user.[6] This is equal to about 570 persons per acre. Now this may be fine for an average, but intuitively we know that most people do not go to the beach for average purposes. Some go to swim, some to sun and socialize, yet others to be quiet at the edge of the sea. That being so, we would be inclined to agree with McConnell and Weaver who say that "—net benefits from salt water beach recreation will be greatest if beach capacities are planned to recognize the diversity of recreation experiences sought by salt water beach users".[6]

Based upon a study of southern Rhode Island beaches, they suggest the following capacity ranges, depending on the intended use of the beach:

Type of beach	Maximum persons per acre
Family beach	250—350
Natural beach in undeveloped area	100—200
Highly developed singles/surfing beach	800—3500

(From McConnell, K. E. and Weaver, T., *Setting Capacity Standards for Salt Water Beaches, Bull. 426*, R.I. Agric. Exp. Stn., Kingston, R.I., 1979, 21.)

The resource requirements other than parking that are most suitable for the various types of beach are, briefly:

Family beach — Gently sloping, fairly wide beach, little wave action or undertow. Service facilities are needed.

Natural beach — The beach can be narrower with a steep drop-off; rocks, coarse sand, and lower water temperatures are tolerated. Minimum to no service facilities.

Highly developed beach — Broad beach, frequent surf, fine sand, a spectrum of commercial service activities close by.

The exact characteristics will vary some, depending on which area in the U.S. one is dealing with. But neither these nor the exactness of the figures are the important points here. What is important is that resource planning decisions affect the nature of the recreational "product" that can be offered, that recreation is very important in our lives, and that, therefore, frustrations caused by lack of well-planned resource allocation in the recreational area carries particularly high penalties.

Since the focus of this volume is on the human aspects of using the seas and coasts, we shall not go further into the natural resource requirements except to note that if the supply of suitable natural resources falls seriously short of the amount needed to yield the *anticipated* recreational experience for those seeking it, then limited access or rationing through price become the alternatives to a destruction of the recreational experience.

Some forms of recreation such as deep-water cruising or racing are naturally price or income-limited, for they are expensive, but in the case of the various forms of recreation on beaches, we have typically rejected the setting of limits and let deterioration of the recreational experience lower the numbers of "seekers" to achieve not a balance but an

Table 3
EXAMPLES OF SOCIO-ECONOMIC DATA ON BOAT OWNERS AND PARTICIPANTS[a]

	Rhode Island 1973[b]	U.S. 1973 Boat operators[c]	U.S. 1973 Population as a whole	Oregon 1977[d]
Household income			(%)	
Less than $10,000	24.4	28.5	44.2	22.4
$10,000 — $14,999	31	32.9	22	17.6
$15,000 — $24,999	29.4	27	24	37.5
$25,000 or over	15.2	11.6	9.9	22.5
Average household income	< $15,000	< $15,000	$12,500[e]	$16,455
Age of primary participant	44.8	34.2	28.2[e]	54.1
Occupation			(%)	
Professional, managerial, administrative	35			17.8
Clerical and sales	9			5.6
Craftsmen, operatives	34			26.9
Proprietors, others	11			16.5
Retired	11			33.2
Family size	3.4			2.1

[a] In comparing the three boating studies one should be aware that, for the country as a whole, household income was about 30% greater in 1977 than it was in 1973.
[b] Data taken from Rorholm, N., Boats and Their People, University of Rhode Island, Marine Technical Report 52, Kingston, R.I., 1976.
[c] Data taken from U.S. Department of Transportation, USCG Report No. 745103, Table 38, Washington, D.C., 1974.
[d] Data taken from Vars, C. R., Jr., Recreational Boating in Western Oregon, Oregon State University, Sea Grant College Program, Publication No. T-79-005, Corvallis, Ore., 1979. (Data are for those boating in salt water.)
[e] Interpolations from Statistical Abstract 1979 U.S. Department of Commerce, 1979.

uneasy truce between supply and demand. In the opinion of this author and of some who have given the matter a lot of thought, this is the wrong solution.[7]

There is no more reason to balance supply and demand for recreation by giving frustration instead of mental and physical renewal than there would be for stretching supply and limiting demand for automobiles by selling some cars without wheels, others with wheels but no doors, etc. To be sure, the market could be brought into balance but no one would be satisfied.

To indicate the type of information that is being made available in most cases on a statewide basis, we shall show examples of research that indicate socio-economic characteristics of participants, some partial data on expenditures incurred by participants, and some estimates of the resulting commercial sector.

VI. PARTICIPANTS AND INCOMES

Table 3 shows profiles from three studies of boat owners (the two state studies) and boating participants including owners (the U.S. study). With respect to income, one can say that a family participating in boating (column 2) was much less likely to have income under $10,000

Table 4
CHARACTERISTICS OF SALT WATER ANGLERS BY PLACE OF FISHING, RHODE ISLAND 1978

	Fixed structure	Shore	Private boat	Party/charter boat
Average age	37.4	38.9	40.1	35.1
Fishing experience, years	17.6	18.8	21	17.1
Household income	\$15,600	\$15,500	\$19,800	\$21,300

in 1974 than was a family that had not owned a boat (column 3), slightly more likely to have income over \$25,000 and much more likely to have income between \$10,000 and \$25,000. In other words, as expected, few with very low incomes own boats but beyond that boat ownership is not, as many assume, associated with high incomes. Studies show boat owners grouped in the two income classes bracketing the median family income level. See also King and Nitschke[8] and Brown and Noden.[9]

Average age of the primary participant ranges from a low of 44.8 years in Rhode Island (column 1) to a high of 54.1 years for Oregon salt water boaters (column 4) in the studies examined.

There are enough differences in the distribution of boaters among occupational categories to thwart an effort to draw a neat picture of an "average boat owner". Note the considerable differences in the distribution in columns 1 and 4, Table 3, particularly in the "professional-managerial" and "retired" groupings. This, when supported by the family size data (3.4 compared to 2.1), describe two quite different populations.

In McConnell et al.,[10] characteristics are listed of Rhode Islanders who fished in salt water in 1978 (Table 4).

It is interesting to note that those who fish from party or charter boats are younger, have less experience, and earn more money than people fishing from other vantage points, even from private boats.

In a study of charter fishing along the Texas coast, it was found that the median income of participants was a little over \$30,000 and that their age averaged 45 years, a little older than the combined party-charter boat population in Rhode Island.[11]

One would expect a distinct difference between incomes of those who frequent party or "head" boats and those who use charter boats. The former leave on regular schedules, can hold many fishermen, and typically anchor offshore while the customers fish with bottom gear; the latter go out with a small group and most often seek gamefish of one species or another — a much more expensive pastime.

The above are examples of the kind of socio-economic information available. By looking at data from the overall survey of outdoor recreation, one can observe the more general relationship between income and participation in selected marine recreation activities. National data for 1977 will be used.

Table 5 demonstrates first of all the overwhelming leadership in popularity of swimming (not in pools) and fishing among water-based recreation, followed by other boating, water skiing, and sailing (column 1) (see also Figures 5 and 6). Further, the lower two income groups had slightly more allegiance to fishing (35 and 41%) than to swimming, while in the case of the higher two groups, swimming came first with 42 and 43%.

Direct comparisons may not be made horizontally in Table 5, for the distributions are computed within income groups, not within recreational groupings. However, the columns still do not add up to 100%, for some people participate in more than one kind of recreation and some people in none at all. Take other boating, for example, on line 3, for which we

Table 5
PARTICIPATION RATE AND PARTICIPATION BY INCOME GROUPS IN SELECTED WATER RECREATION, 1977

	Percent of 4,029 person sample participating[a]	Percent of 3,310 persons in income groups participating in selected recreations[b] (thousands of dollars)			
		10 or less	10.1—15	15.1—25	25.1/over
		(percent)			
Swimming	35	27	38	42	43
Sailing	5	3	2	5	14
Other boating	20	15	21	24	30
Water skiing	8	4	10	9	14
Fishing	36	35	41	37	35
Percent of 3,310 sample in specified family income groups		31	24	30	15

[a] Those who participated five or more times in 1977.
[b] Those who participated and also declared their income.

(From Third Nationwide Outdoor Recreation Plan, Appendix 2, Heritage and Conservation Service, U.S. Department of Interior, Washington, D.C., 1979, 19.)

FIGURES 5 and 6. Swimming and fishing are the leaders in marine recreation.

FIGURE 6.

have quoted more detailed studies that indicate income of boaters is typically somewhere in the middle rather than very high. It is tempting to read line 3 as saying that twice as many people with over $25,000 income went boating (30%) as did those with $10,000 or less income (15%). But wait, line 6 shows that 31% of 3,310 people were in the first income group (1,026 people), 15% of those went boating five or more times (line 3), 154 people in the sample. Fifteen percent of the sample had the higher incomes (509 people), 30% of them went boating five or more times (153 people). So almost an identical number of boaters had high incomes as had low incomes.

The proper interpretation of distributions such as this can be significant. For example, if

Table 6
INCOME DISTRIBUTION OF PARTICIPANTS IN SELECTED KINDS OF WATER BASED RECREATION 1977

	Percent of participants in outdoor recreation activities having specified incomes (thousands of dollars)				
	10/less	10.1—15	15.1—25	25.1/over	Total
			(per cent)		
Swimming	23 [48]	25	34 [52]	18	100
Sailing	19 [29]	10	29 [71]	42	100
Other boating	22 [45]	23	33 [55]	22	100
Water skiing	15 [43]	28	32 [57]	25	100
Fishing	28 [55]	27	30 [45]	15	100
U.S. family income	27 [46]	19	32 [54]	22	100

(From Table 5 and Statistical Abstract of the United States, U.S. Department of Commerce, Bureau of the Census, 1979, 448.)

questions of user fees are to be discussed by policy makers, then it becomes very important that the data show income distribution within uses. In line 1, for example, it could appear that a high proportion of swimmers (43%) had high incomes and could afford substantial user fees. Let us convert Table 5 to one showing income distribution within uses. We will facilitate comparison by keeping row and column headings in the same order.

Table 6 has the answer: 59% of swimmers have income from $10 to 25,000, and a greater percentage (23) has lower than has higher incomes (18). Therefore, to conclude there is wealth among average beach goers that would enable them to pay high fees would be incorrect.

Table 6 can also facilitate a coarse comparison of our recreators with the income characteristics of the population as a whole. First, we note that in all cases except sailing, the highest proportion of households had incomes in the $15 to 25,000 group (as is the case for the population as a whole). Sailing appears to differ significantly from the other groups in the proportion of higher incomes of its participants. We can get a simpler look at this by considering only two groups, lower ($15,000 or less) and higher (over $15,000). Those numbers are given in square brackets in Table 6. Studying them leads to the following generalization: on the average, it appears that people who participate in swimming, other boating, and water skiing have incomes typical of the general population, whereas the average sailor has a significantly higher income. There are somewhat more people in lower income groups among those who fish for recreation than is the case for the population as a whole.

From his experiences studying boating populations, the author would judge that the higher average income found in sailing is a manifestation, not so much of traditional or concentrated wealth as of the fact that sailing is a part of the active outdoor recreation increasingly being sought by successful younger families, many of them with two income earners.

VII. PAYING THE WAY

We have alluded to the great variation in what people pay for their marine recreation. The land-based resources being used can be in either private or public ownership, the water is free to the user, and equipment costs vary from the cost of a string and a hook to the investment in a large house or a yacht or both. In addition to the costs borne by participants, the public lands must be kept up, buoyage maintained, and occasional rescue operations carried out. These are the public costs of marine recreation and they are covered either through taxes or specific fees; more often than not a combination of the two is used.

Studies have been made of the expenditures incurred for some kinds of recreation activities. One commissioned by Boating Magazine identified 7,010,000 boat owners in the U.S. of which 3,355,000 or 48% had boats under 16 ft long; 3,393,000 or 48% owned 16 to 24 ft boats; and 592,000 or 4% had boats 25 feet or longer.[12]

Average annual costs of owning and operating boats in the three size classes based on studies in 1973 in New York and Rhode Island was $176, $494, and $1,730, respectively.[13]

Considering that the Consumer Price Index was 133.1 in 1973 and 214 in 1979, one may estimate the annual expense per boat in 1979 as:

Under 16 ft — $283
16 to 25 ft — $795
26 ft & over — $2,784

Multiplying these numbers by the numbers of craft determined by the Boating Magazine Survey yields an estimated annual expenditure of $5.3 billion in 1979, not counting interest on investment. Other methods will yield other numbers. For example, Boating Industry Magazine estimates that there are 11.8 million boats in the U.S. and that annual sales of supplies, services, and new boats total $7.4 billion.[14]* Whichever source is used, the amount is substantial. Considering that two years separate the studies and the latter numbers include new boats, it is surprising the numbers are not farther apart.

Expenditures of those who fish for recreation have been estimated also. While, again, costs vary greatly, a study conducted in 1970 indicated that for the country as a whole about 706.6 million fishing trips were taken at an average cost of $7.02/angler/day for a total expenditure of $4.96 billion.[15] Considering the 89% increase in the Consumer Price Index from 1970, the total expenditure in 1979 would have been about $9.38 billion, or $13.27/angler/day assuming no change in the number of anglers, a very conservative assumption.

It is, in fact, very difficult to separate the expenditures from boating and fishing, for recreational boats are used a large proportion of their time for fishing. Table 7 shows this as well as a great variation in use patterns, depending on the geographical area involved.

In an attempt at summing up the expenditures on marine recreation, one might look at the amount of business generated by the spending. The sources for Table 8 are numerous Bureau of the Census reports and a survey of the nation's Sea Grant programs in 1979.

The data indicate that more manpower and firms are dedicated to supporting marine recreation than to either of the other two important uses of the marine environment. In total sales, marine recreation is second to fisheries. The data are, of course, incomplete for reasons we have discussed elsewhere and because of the different reporting practices in the fishing sector. In addition, some double-counting of total sales is inherent in this use of secondary data. It can only be eliminated by the U.S. Department of Commerce's recording of commercial sectors, as suggested by Pontecorvo.[16] It is the judgment of the author that the elimination of double-counting would affect transportation, fisheries, and recreation in that order, leaving recreation in a leading commercial position among the three if a complete accounting were possible.

Do the participants get value for their money? It would appear so. Florida residents placed an average value of $63/year in 1971 on their right to free beach access.[17] If that is the value received, it is certainly more than had to be paid for beach access at the public beaches

* Compare to seven-plus million boats identified in Reference 12. Counting boats, is, at best, an inexact science. Whether measured through registrations, manufacturer's shipments, or dealer data, there is a tendency to end up with overestimates. Then there are problems with definitions. Are boats without mechanical power included, for example?

Table 7
PERCENT DISTRIBUTION OF BOAT TIME IN THREE MARINE AREAS

Category	Oregon	New York	Rhode Island
Fishing	91	40	49
Water skiing	1	7	6
Day trips/cruising	4	45	37
Other	4	8	8
Total	100	100	100

(From Table 3 and Reference 9.)

Table 8
SELECTED MEASURES OF MARINE COMMERCIAL ACTIVITIES BY FUNCTIONAL CATEGORY, 1979

Category	Employment	No. of firms	Payroll ($1,000)	Total sales ($1,000)
Fisheries	355,669	7,025[a]	2,459,755[a]	27,850,368
Marine transportation	351,749	3,646	5,908,076	15,666,618
Marine recreation	806,143	42,297	4,867,440	23,552,971

[a] Catching sector not included.

that did charge for parking or admission at that time. Horvath found that salt water anglers valued a day of fishing at about $60 in 1973 and about $112 in 1979 prices.[18] If this is so and if their expenditures were as reported above, the fishermen are clearly ahead of the game.

The issue of costs compared to value received is an important one, particularly in cases of activities in publicly-owned recreation areas. The recreational experience is a composite that is not easily studied in commercial markets. It is therefore difficult to know what value these areas have when used for recreation as opposed to some commercial use. To facilitate the estimation of value to society of using an area for recreation or preservation as compared to the market values of commercial use, economists use various analyses of willingness-to-pay.

VIII. CONFLICTS IN MARINE RECREATION

Whether marine recreation will continue to grow and whether that growth will benefit the participants will depend a great deal on our ability to formulate and carry out reasonable policies for coastal use.

Any use of land and water resources carries with it the potential for conflict with other uses (Figure 7). The conventional wisdom has it that the typical conflict occurs between recreational and commercial uses. In our mind's eye we see factories, junk yards, and ports pushing along a shoreline that might otherwise be used for recreation. But, observation tells us that this picture is false. The 70% of the U.S. coastline (excluding Alaska) which is in private ownership, is not predominantly in industrial/commercial or port use. It is in residential or recreational/commercial use. It is providing recreation.

We need to elaborate on two aspects of this generalization. First, ports and industry. It

FIGURE 7. Any use of water and land resources carries with it the potential for conflict with other uses.

is certainly true that many stretches of coastline exist in urban areas that are no longer effectively used for their original commercial purposes and which could well provide fishing spots, picnic areas, boat launching sites, promenades, restaurant space, even marinas for recreational purposes. The slow pace at which these needs are being met in urban areas is frequently caused by the land's being held for speculative purposes, the low rent that can be afforded by anything but luxury-type recreational developments, and the taxation practices of the cities. But progress is being made, for example, in Savannah and Seattle, and no doubt in many other cities.

The second aspect of the generalization we need to look at is the nature of the recreation provided by the privately-held coastline. It is dictated to a large degree by market forces and we will illustrate those with another generalization: in temperate areas where relatively short seasons are the rule at the shore, a resort complex cannot afford to bid against the four or five wealthy individuals who want to share among them the same prime beach area for their residential purposes. Also, in the more southern climates it takes a great deal of wealth to bid away from a hotel or motel the needed area of prime beach for residential purposes.

Zoning has been used to regulate the mix of private and public/commercial marine recreation but market forces are difficult to counter without ownership. Thus, it is fair to say that without public ownership there would be no reasonably priced marine recreation.

In protected coastal areas where boat service facilities might be located we find that competition for space frequently comes from either residences or natural areas (marshes), which typically form there. The income potential of the marina precludes competition with private residences and the natural areas are expensive to develop even if it were possible or desirable.

The competition for space, then, in the vast majority of cases, is with the same form or with other forms of marine recreation, not with commerce and industry.

Another type of conflict, which we might call environmental, when imposed upon a commercial use renders it less profitable; imposed on a private use it renders it less enjoyable. Commercial examples are oil spilled on a hotel beach, commercial gill netting in favorite fishing spots for party boats; private examples are other boats snaring one's fishing lines on a favorite fishing spot and crowded anchorages making cruising difficult.

This aspect of conflict is more likely to have a commercial/industrial source, but in looking at reports of these instances over many years the author concludes that even here the great majority are caused by residences (sewage) or by other forms of recreation. It would appear that when it comes to situations that prevent us from getting optimum use of our marine recreational resources, we are indeed our own worst enemy.

IX. THE FUTURE

As we look toward the end of this century, we see marine recreation activities adapting to two important factors: stagnant or possible decreasing real disposable incomes, and continued high energy costs.

If we think first of the spatial distribution of marine recreation, we would expect pressures to intensify close to the population centers. We would expect to see people become less willing to drive far to reach prime recreational areas. They will probably satisfy themselves with a less pleasurable experience and will, in some instances, substitute cheaper means of recreation: buy smaller boats perhaps, share the ownership of large boats, fish from shore instead of from a boat, swim at beaches close to home instead of traveling to resort areas. However, if foreign travel is also cut back, there may not be a significant reduction in resort business. Similarly, there will be enough boating people with sufficient income to maintain their boats at marinas, although they may use power boats somewhat less. But it will be a "cold day" before the dedicated sailor of even adequate means folds his wings in response to the income/cost squeeze.

From the point of view of recreational space, then, we would expect the pressure to return to the coasts close to cities, and that includes parts of the cities. It is, therefore, important that this potential pressure be recognized and outlets found through means such as public transportation facilities to beaches, urban waterfront parks, fishing piers, urban small-boat sailing clubs, and rowing clubs, to mention a few possibilities.

Will the resources hold up to the pressure? Except in one instance, the resources are space on land and water, capital, and manpower. If they earn adequate returns, they will not be switched to other uses. The land will not wear out although it may move around a bit. So, except for the one case, there seems to be little danger that we will run out of resources for marine recreation. It is far more likely that crowding will so change the experience that many people will turn to other activities.

The one exception in this resource prediction is marine recreational fishing. Sportfishing, as such, will not empty the ocean of fish, but recent data indicate that the recreational catch is considerable, and if quotas or other restrictions are placed on the commercial fleet, will it be long before limits are placed on recreational fishermen? To put the magnitude of the catch in perspective, it is instructive to look at the results of the 1979 Marine Recreational Fishing Survey of the Gulf and Atlantic Coasts.[19] The data indicate that 9.2 million anglers caught 298 million fish in 58.8 million trips. The total weight landed was 346.8 million lbs or about the same weight as all the flounder, cod, haddock, red and white hake landed by American commercial fishermen in the Gulf and Atlantic. Another way to look at it is that assuming an efficient modern 80-ft stern trawler with a five-man crew catches 1.25 million lb of foodfish per year, it would take 277 such trawlers to catch the equivalent of the reported recreational catch. It would not be surprising if this heretofore fairly low-key issue of the sharing of the catch were to become a difficult marine resource management issue in the future. As pointed out by McConnell et al.,[10] because of the high catch rates from boats as compared to fishing from shore or fixed objects, it would do little good to merely regulate the fishing from shore. Here again, as with crowding, there is always the option (not recommended) of letting deterioration of the recreational experience lower the demand, although with both commercial and sportfishermen seeking the same fish, the situation is a bit more complicated.

People will continue to seek the sea for recreation. Since the shore is the meeting of public and private domains and inasmuch as both sea and shore are needed as recreational resources, there must always be some degree of involvement of public policy bodies in the allocation of resources to recreation and among types of marine recreation. The market mechanism can and will do much of it. But there remains a significant need for social, economic, and technical research to assist the decision makers in this resource allocation process.

REFERENCES

1. **Gold, S. M.,** User needs and requirements in recreational access to the coastal zone, Proc. March 1978 Forum, Fawcett, J. A., Sorensen, J. C., and Manus, A. T., eds., University of Southern California Sea Grant Program, 1978.
2. The Third Outdoor Recreation Plan, U.S. Department of the Interior, Heritage Conservation and Recreation Service, Washington, D.C., 1979.
3. Sea Grant Situation Statement and Recommendations on Marine Recreation, Report from the 2nd National MAS Recreation Workshop, Norfolk, Va., March 27-28, 1978.
4. **West, N.,** Coastal demographic changes in the United States, 1950-1974, Proceedings of Oceans-78, Marine Technology Society, Institute of Electric and Electronic Engineers, MTS Annual Conf., Washington, D.C., Sept. 1978.
5. **Rood, M. and Warren, R.,** *The Urban Marina, Managing and Developing Marina Del Rey,* University of Southern California, Los Angeles, 1974, 57.
6. **McConnell, K. E. and Weaver, T.,** *Setting Capacity Standards for Salt Water Beaches, Bull. 426,* R.I. Agric. Exper. Stn., Kingston, R.I., 1979, 21.
7. **Meyer, P. A.,** Recreational policy-resource evaluation for decision making, in *Recreation: Marine Promise,* Anderson, S. H., Ed., University of Southern California Sea Grant Program, Los Angeles, 1977, 193.
8. **King, F. and Nitschke, D.,** Maine's pleasure boat industry, Part II: Maine boaters, *Bull. 733, Life Sciences and Agricultural Experiment Station,* University of Maine, Orono, Me., 1976.
9. **Brown, T. and Nolen, D.,** New York Recreational Boating Survey, Department of Natural Resources, Cornell University, Ithaca, N.Y., 1977.
10. **McConnell, T., Farrell, J. F., and Smith, T. P.,** Marine Sportfishing in Rhode Island, 1978, URI Sea Grant Marine Technical Report No. 83, Kingston, R.I., 1981.
11. **Mertens, T. J.,** Charter Boat Fishermen on the Texas Coast, Texas A&M, Sea Grant College Program, College Station, Tx., 1977, 7.
12. Nationwide boat owner survey, a study of boat ownership, purchasing and reading behavior, *Boating Magazine,* Ziff-Davis Publishing Co., 1977.
13. **Rorholm, N.,** unpublished data, 1973.
14. *The Boating Business,* Annual Business Issue, January 1981.
15. National Survey of Hunting and Fishing, 1970, Department of the Interior, U.S. Fish and Wildlife Services, Resource Publication 95, U.S. Government Printing Office, Washington, D.C., 1972.
16. **Pontecorvo, G.,** Contribution of the ocean sector to the U.S. economy, *Science,* 208, May 1980.
17. A Report on the Sea Grant Program, State University System of Florida, Gainesville, Fla., 1972.
18. **Horvath, J. C.,** Southeastern Economic Survey of Wildlife Recreation, Environmental Research Group, Georgia State University, Atlanta, Ga., 1974.
19. Fisheries of the United States, 1980, Current Fisheries Statistics No. 8100, NOAA Department of Commerce, Washington, D.C., 1981.

Chapter 2

FISHING PEOPLES

Courtland L. Smith and William Q. Wick

TABLE OF CONTENTS

I. INTRODUCTION

Anthropologists emphasize two major processes when studying how cultures change — evolution and acculturation. The process by which cultures change in form through time is called evolution. The shift from capturing to culturing fish is an example. Fish capture is associated with cultures that have small, egalitarian communities, that have a less formalized social and political structure, and where economic growth and development are not emphasized as goals. Usually evolutionary changes occur independent of contact with other cultures.

The process of change resulting from contact between cultures is acculturation. It is generally viewed as change that comes from forces external to a culture. For example, when settlers entered the Pacific Northwest, they learned dipnetting and techniques for drying salmon from the Indians. Settlers incorporated these techniques with ways of their own for catching and preserving salmon. Acculturation changed Indian culture more by introducing concepts about property rights, participatory democracy, and domestication of resources. Indian cultures were interred on reservations, encouraged to formalize tribal governments, and advised to give up fishing for farming. Reservations established the idea of private property. They separated the land allotted Indians from the farms and ranches of settlers. Indian leaders were selected for tribal government where decisions were based on majority rule, rather than the patterns of concensus that had ruled in the past. Most difficult for Indians of the northwest to understand was the white orientation to nature. Her bosom was ripped with plows, her waters diverted and polluted, and her fish runs stopped by great concrete and earthen barriers.

The salmon fishing peoples of the Pacific Northwest, before the coming of whites, depended on "one, or a few, local ecosystems".[1] These "ecosystem" people, and others like them, saw themselves as living in harmony with nature. They viewed the earth as their mother. When asked during treaty negotiations to sell their land, they said, "We cannot sell the mother earth".[2,3]

Explorers, traders, and settlers were "biosphere" people. They organized trade systems to exchange resources between ecosystems.[1] They did not, however, think of the earth as one biosystem. When Robert Gray came to the Columbia, he was trading cloth, brass buttons, and chisels with Native Americans for pelts which he took to China in trade for tea. Bostonian merchants developed great wealth not from their productive genius, but from their Yankee ingenuity for exchanging goods between ecosystems.

Communication advances bring more interaction between the world's ecosystems, and enable biosphere societies to extend their technology and economic arrangements. This allows exchanging items and experience between ecosystems but it also threatens the existence of ecosystem cultures.

II. OVERVIEW

The dual processes of evolution and acculturation point to the extinction of fishing peoples as we have known them in the past. The pattern of evolutionary change from capture to culture means fishing peoples are less able to follow traditional pursuits. As ecosystems are modified to accommodate practices associated with agriculture and aquaculture, the shifting, relatively egalitarian, and less formal fishing peoples give way to dominating, sedentary, stratified, and more formalized cultures.

This coupled with the process of acculturation where dominating economic and political systems link ecosystems together, too operates against persistence of fishing peoples. Economic development involves implementing practices developed in one ecosystem, and applying them elsewhere. This fails to recognize the interrelatedness of cultural and ecological systems. Biosphere peoples seek to exchange ideas and goods between ecosystems. This makes local ecosystems more dependent on events beyond their boundaries. The resources from which fishing peoples gain their livelihood become subject to the fluctuations and idiosyncracies of distant markets.

One perspective on the evolution of human society is that fishing peoples are an offshoot from the line of human evolution.[4] An alternate view, and the one taken here, is that aquatic resources are important to some degree in most cultures. Fishing peoples are those cultures where an aquatic orientation is integrated into their basic way of life. This can be either marine, riverine, or lacustrine. With few exceptions, most humans, until the last several millennia, have been rather timid about venturing away from land. Trans-Atlantic and Pacific migrations may have occurred as early as 5000 years ago.[5] Until the 14th century, occidental culture knew only 7% of the world's oceans and 21% of the land surface.[6] Columbus was one of the first Europeans known to have left sight of land. The Nina, Pinta, and Santa Maria were all less than 30 m long.

Polynesians, more than a millenia before Columbus, mastered overseas colonization using rafts.[7] De Bisschop[8] labels Europe's explorers as "navigators". A navigator "dislikes being out of sight of land. He travels only for practical reasons; commercial ventures, conquest, the abduction of slaves." Search for the Northwest Passage powered much European ocean exploration. By contrast, de Bisschop characterizes the Polynesians as mariners, people who go to sea "in a spirit of adventure and discovery". More critical than the spirit of the explorers is their notion of time. Occidental cultures see time as a linear process that regulates the occurrence of events. For the Polynesian, time merges together life as a whole.[8]

The land-sea interface affected human development much more than ocean exploration. Sauer[9] advances the idea that the shoreline may have been the divergent path of human evolution and expansion. Here at the interface between land and sea the resource pool is rich, varied, and easily obtained. Following the shore, human expansion to new territories is possible without major adjustments in adaptation. Early *Homo sapiens*, according to Sauer, kept to the coast and did not venture out into the oceans. Due to varying sea levels over the last million years, much of the evidence to test Sauer's hypothesis is lost to the archaeological record.

Humans are very versatile. They do not limit themselves to any one resource. Fish, seaweed, and marine mammals all contribute to the diet of some cultures. Along the Northwest coast, Makah whalers captured migrating whales from their seagoing canoes (Figure 1). Eskimos developed a complex technology to take seals and sea lions. In fact, all the fishing techniques used today, e.g., seines, traps, longlines, trolling, and nets were used aboriginally to catch fish.

The use of aquatic resources followed very much the same pattern as terrestrial resources. Capture led to culture. Perhaps the women gatherers who may have started plant domestication, also started aquaculture. Who knows? The answer is lost in Chinese prehistory. We do know that more than 3000 years ago humans began the process of domesticating aquatic animals. It is likely the first step in this process was similar to the capture of milkfish. They are taken at sea as fingerlings and reared in brackish coastal ponds. The Japanese were the primary inventors of shellfish culture. Culturing salmonids and ocean going trout is a contribution of the early U.S. fishery program to replenish sport fish lost due to polluted East Coast streams.[10]

Since neither fishing nor aquaculture is a dominant occidental cultural tradition and since those of Indo-European extraction seem prouder of their success with bovine, barley, and wheat domestication, fish do not receive the same emphasis that they would in Chinese, Indonesian, or Polynesian culture. Norse, Icelandic, and Irish sea fishers continually remind us that peoples of Indo-European origin turned seaward. But for many, fear characterized their relations with the sea. Bullen[11] writing in 1901 observes, "So many highly educated people among us have a horror of the sea, regarding it as a dread and dreary expanse of

FIGURE 1. Henry Wood Elliot drawing of Makah whalers, 1883. Courtesy of Robert L. Hacker.

heaving billows, concealing terrors greater and dangers more immense than even the ancients dreamed of . . . "

Europeans became biosphere people, searching for better trade routes to augment a rather boring and drab existence. To do this they had to master the ocean ecosystem. Cook proved that travel at sea need not be followed by death and disease. By carefully planning the nutrition of his crew, he prevented the large number of deaths usually resulting from scurvy in a long ocean voyage. Where 221 of Magellan's 268 member crew died, Cook's Resolution crew of 113 lost only four in three years of exploration.[12] Further, only one of these deaths occurred due to the rigors of ocean travel. Improved diet was mainly the reason.

A strong fishing tradition exists in most European and North American coastal communities. Fishing, however, is one among many economic activities. It is not normally dominating. Usually fishermen have to fight to maintain their opportunities in competition with maritime developments focused on other resource exploitation, trade, and national security. At the supermarket meat counter, the place of prominance is given to beef, not fish.

Quite the opposite is true for those with Chinese heritage. Fish is served at all important family, social, and religious occasions. Carp are a common theme in Chinese paintings. "Fish occupy such a high position in the tradition of Chinese history that fish culture developed in China widely and extensively."[13]

Prior to evolution to fish culture and acculturation to exchange products between ecosystems, fishing peoples existed whose culture integrated fish and fishing as part of one cultural and ecological system. Among fishing peoples, a change in fishing behavior affects this entire system. Because of their integration with the ecosystem, evolution and acculturation result in altering both natural and cultural system features.

III. ETHNOHISTORIC PATTERNS

Data from the Human Relations Area Files can be used to establish some of the subsistence

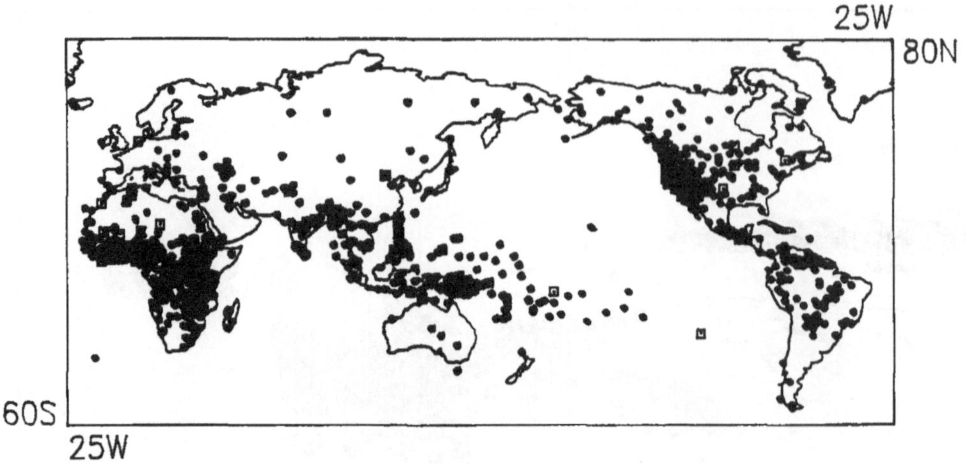

FIGURE 2. Worldwide distribution of the 1170 cultures used for the analysis of fishing patterns. This reflects the cultures studied by anthropologists and their location at the time of contact. Heavily studied areas are associated with European colonization and endeavors to learn about the ways of peoples in these regions.

patterns characterizing human societies. The following maps are drawn using data coded from the *Ethnographic Atlas.*[14] Unfortunately, these coded data do not allow for identification of the specific resources making up the subsistence pattern, but they do show the geographic predominance of several subsistence patterns.

Figure 2 shows the geographic distribution of cultures used for this analysis. The distribution reflects cultural locations at the time of contact between European colonizers and indigenous peoples. The points represent each culture's center of mass located to the nearest degree (110 km at the equator). Cultures spread over large areas are located according to the place where data describing them were gathered. Within a degree, several distinct cultures may be located. While the points plotted reflect neither the geographic area encompassed by a culture, nor the population size, they do show cultural concentrations by geographic area.

For the 1170 cultures plotted, 80% are in the heavily colonized areas of North and South America, Africa, and Oceania. Europe and Asia have only 20%. These are biases of representation, but the data are the most comprehensive source available and are adequate for comparison of regional differences in subsistence patterns.[15,16]

Fish forms some part of the diet for 66% of these cultures. It makes up more than 5% of the diet in nearly 40% of the cultures. Fishing, then, is a pervasive cultural practice, and among cultures in which it is important, there are several interesting fishing patterns.

What do we know about the fishing patterns and peoples of the world? Several summaries approach this question.[17-24] The literature on fishing peoples appears in diverse places. For bibliography see Alkire,[25] Andersen,[26] Drucker,[27] Harding and Wallace,[28] Landberg,[29] Murdock,[30] Murdock and O'Leary,[31] Oliver,[32] Poggie et al.,[33] Steward,[34] and Vayda.[35]

Figure 3 documents the northwest coast of North America concentration of fishing peoples at the time of contact with European explorers and settlers (Figure 4). Salmon were the principal subsistence resource for most of these cultures prior to contact, although other marine resources, too, were important.[27,36,37] Figure 3 plots 71 cultures. These cultures all depended on fishing for 55% or more of their subsistence. Half are northwest coast cultures, and all but eight cultures are located in North America. One third are Eskimo, depending mainly on marine mammals.

Northwest coast cultures illustrate the integration between cultures and ecosystems. Fishing

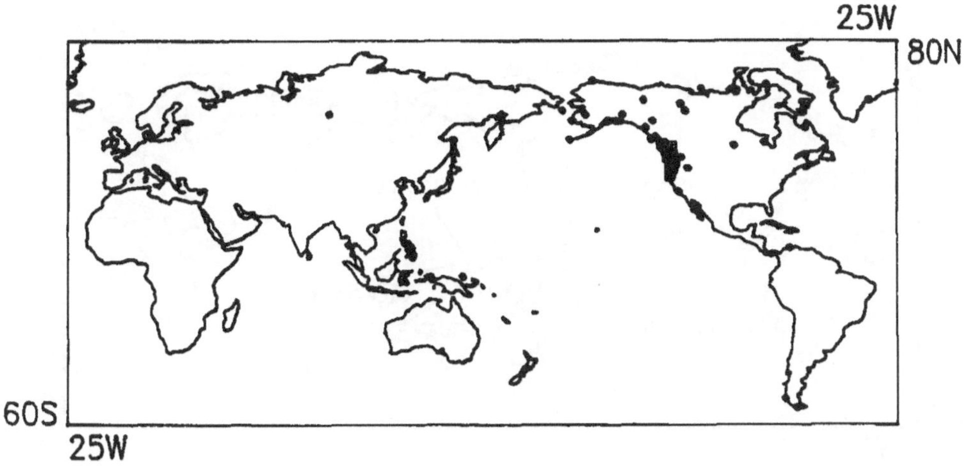

FIGURE 3. Distribution of cultures, showing concentration on the northwest coast of North America of peoples, who depend on fishing for 55% or more of their subsistence and do not practice agriculture.

FIGURE 4. Henry Wood Elliot drawing of a Fraser River salmon trap, circa 1880. Courtesy of Robert L. Hacker.

peoples, like the Kwakiutl (Figure 5), believed that salmon were supernatural beings who live in the ocean as Salmon-people.[27] Annually they sacrifice themselves to sustain human beings. So that the Salmon-people can reflesh themselves, their skeletons are deposited in the water for return to their houses beneath the sea. When caught, northwest coast peoples have special ceremonies and prayers for salmon. A Kwakiutl prayer offers thanks.

We have come to meet alive, Swimmer. Do not feel wrong about what I have done to you, friend Swimmer, for that is the reason why you come that I may spear you, that I may eat you, Supernatural One, you, Long-Life-Giver . . . you have come trying to come to me, you, who always come every year of our world, that you come to set us right that we may be well. Thank you, thank you sincerely, you, Swimmer. I mean this, that you, please, protect me that nothing evil may befall me, Supernatural One, you, Swimmer.[38]

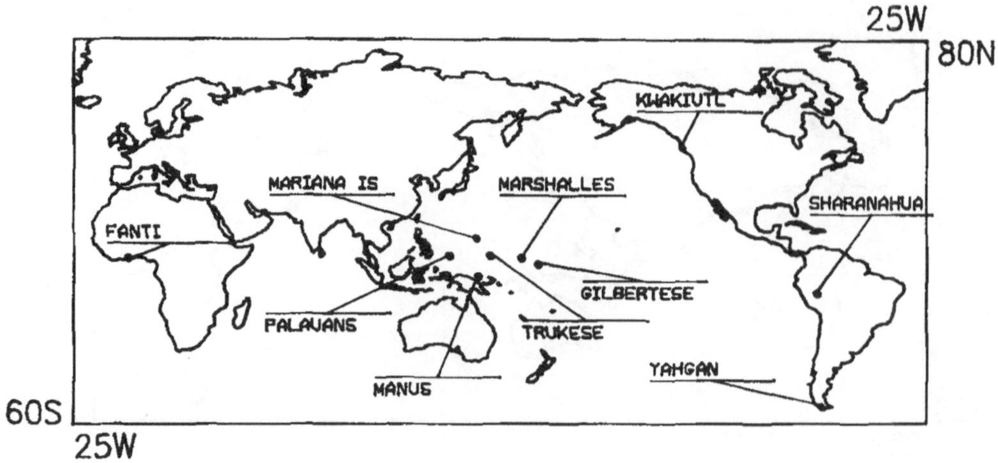

FIGURE 5. Location of cultures illustrating fishing patterns discussed in text.

The peoples with the highest dependence on fishing listed in the Human Relations Area Files are the Manus studied by Mead[39,40] and Fortune.[41] The Manus live in villages of about 200 in the Admiralty Islands. Their houses are built on piles in sheltered lagoons. The Manus fish and trade fish with their neighbors for garden products. Governmental intervention in island affairs reduced intertribal warfare and strengthened the Manus trading position.

Village life was regulated by the requirements of fishing. Each man is driven by a supernatural to hard work that is believed to prevent illness and death. Like northwest coast peoples, the Manus were tightly integrated with the supernatural.

The linkage between fishing peoples and their ecosystem is perhaps best illustrated with extinction of the Yahgan. These people lived in the island archipelago around Cape Horn at the southern tip of South America.[42] The climate is cold, but the major garment worn was a small cape hung over the shoulder and reaching the waist. Sometimes it was omitted entirely in a climate where mean summer temperature was 10°C. Women waded in the cold waters gathering mussels, thought to be a gift from a Supreme Being, without apparent discomfort.

Missionaries insisted on clothing the Yahgan and introducing European ways. Within 50 years their population declined from nearly 3000 to less than 100 due to respiratory diseases, measles, typhoid, whooping cough, and smallpox. Interference with such an apparently insignificant cultural trait as dress, appears largely responsible for the eventual extinction of the Yahgan.

Figure 6 shows the mixed fishing, hunting, and gathering peoples concentrated in the North American Plateau, Great Basin, and coastal California regions. These people shifted between fishing, hunting, and gathering depending on the season of the year and the availability of each subsistence resource. Typically, the subsistence pattern relies in relatively equal proportions on a predominant aquatic, as well as a terrestial plant and animal resource. For Plateau peoples salmon fishing, gathering camas roots, and hunting deer and elk were the main contributors to the subsistence pattern.[43] The Great Basin subsistence pattern exploited contiguous but dissimilar microenvironments.[44,45] Fishing for salmon was important to the groups with access to the Snake River system. Trout was the preferred species in land locked lakes and rivers. Principal plant and animal resources were pinyon nuts, that had highly variable annual yields, Indian rice grass, rabbits, ground squirrels, and pronghorns. For coastal California, shellfish, acorns, and large game (deer and elk) were the dominant elements in the subsistence pattern.

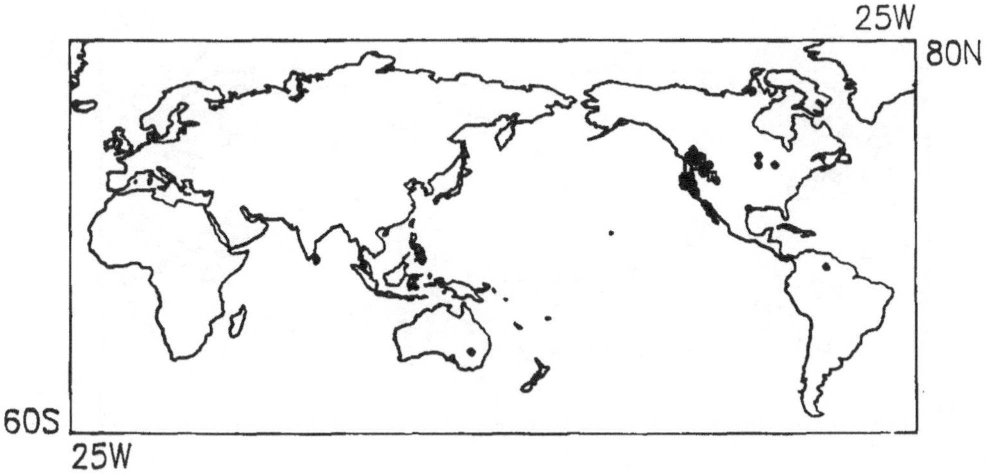

FIGURE 6. Distribution of cultures, showing the concentration in the Great Basin and Coastal California of peoples, whose food comes in roughly equal proportions from fishing, hunting, and gathering.

These data, showing the western North America concentration of fishing peoples, are facts at the time of contact. For eastern North America, Snow[46] documents a fishing-caribou hunting complex of 5000 years ago. As caribou declined they were replaced by moose. European contact brought fur trade and the typical population decline resulting from disease and starvation. As population was lost, so was knowledge of precontact cultures.

Settlers, explorers, and ethnographers in western North America preserved accounts of these peoples by recording their observations of these cultures. These ethnographic data are preserved in the Human Relations Area Files from which the maps were made.

Coastal Chile, too, had maritime oriented cultures as early as 7000 years ago.[47] Six thousand years ago agriculturalists turned more toward the sea. Fish, shellfish, and ocean birds made up the major part of their diet. Then, 2500 years ago, horticulture stimulated greater use of the interior, and domestication of the llama and maize moved the center of cultural development inland.

Moseley[48] poses the revolutionary hypothesis that Andean civilization was not rooted in plant domestication, but maritime adaptation. Moseley documents a shift along the Peruvian coast from fishing to farming. Early settlements were located at the mouths of streams that cut through the coastal desert. Farmers shifted inland from the coast when the demands of inland irrigated agriculture necessitated movement upstream where steeper gradients made canals shorter, facilitated water control, and were less likely to become water-logged. Further, Andean civilization flourished at a faster rate because sedentary coastal peoples already had the agricultural experience and social institutions to mobilize large work forces. Maritime cultures were preadapted, according to Moseley, to assume the requirements of complex irrigation systems. Maritime bases for middle American cultural evolution are summarized in Stack and Voorhies.[49]

A. Domestication

Relatively few plants and animals have become important domesticated resources. Wheat and barley highlight the plants domesticated in southwest Asia. Maize and manioc are the most important New World domesticated plants. In south and east Asia rice is the key plant. For Melanesia and Polynesia yams and taro predominate, while millet dominates Africa. Cattle, sheep, and goats were domesticated in southwest Asia and spread to Africa and Asia. Pigs and chickens characterize southeast Asia and Oceania. Chinese carps, African tilapia, North American salmon, Japanese oysters, and mussels dominate the aquatic animals.

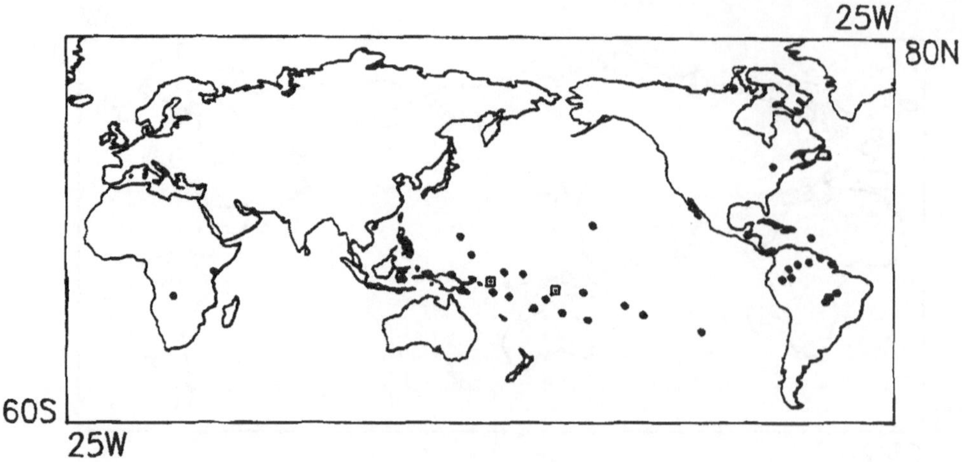

FIGURE 7. Distribution of cultures, showing the concentration in South America and on the low islands in Oceania of peoples, who obtain subsistence from fishing and horticulture.

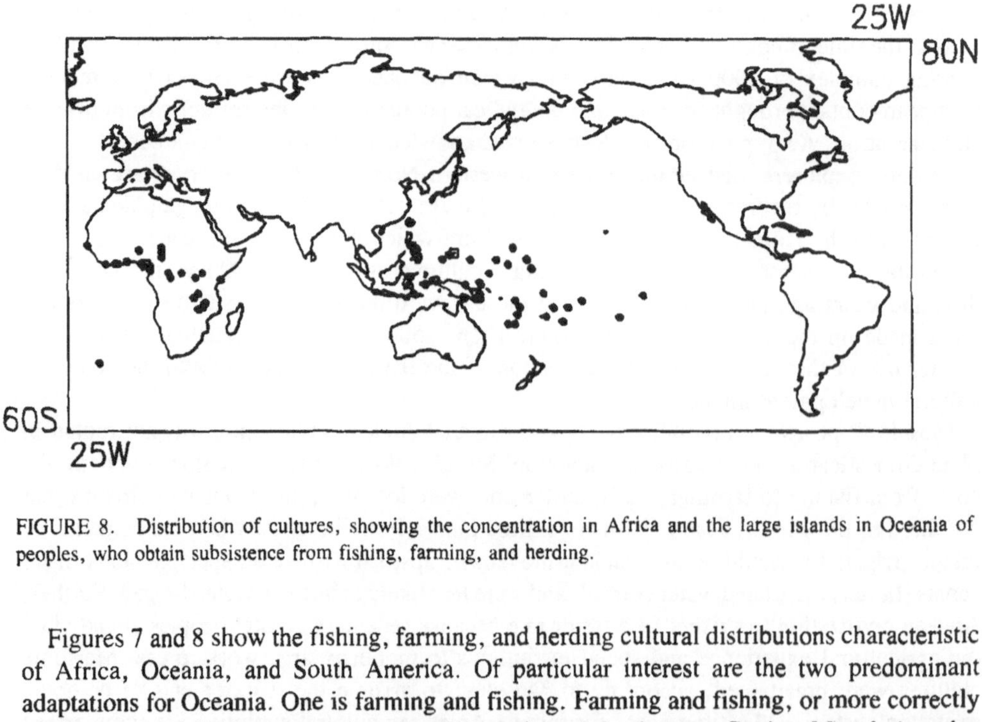

FIGURE 8. Distribution of cultures, showing the concentration in Africa and the large islands in Oceania of peoples, who obtain subsistence from fishing, farming, and herding.

Figures 7 and 8 show the fishing, farming, and herding cultural distributions characteristic of Africa, Oceania, and South America. Of particular interest are the two predominant adaptations for Oceania. One is farming and fishing. Farming and fishing, or more correctly horticulture and fishing, occur in parts of Oceania and the Amazon Basin of South America (see Figure 7). Cultures with this subsistence pattern tend to be relatively small tribal societies. They practice slash and burn horticulture. In Oceania, they occupy low island atolls. In the Amazon Basin they are indigenous peoples depending on a mixture of subsistence activities.

The Sharanahua (see Figure 5) illustrate one of the Amazonia groups who fish by drugging their prey.[50] Men and boys spear the fish as they surface. While returning to the village by canoe, women clean, scale, and wash the fish. Again the supernatural intervenes. Menstruating women or men who have had sex, if they fish, will make the fish immune to the drug.

Horticultural work, too, is a joint activity with men and women complementing one another. Men clear the gardens and plant manioc in the soil beneath the ashes. Maize and squashes are planted in the same plots. Women weed and harvest the garden crops. Hunting also supplements the Sharanahua diet. The Sharanahua village has about 100 people.

Pacific Islands vary greatly in size and resources.[32] Treeless atolls and coral islands, like Johnston, could not sustain permanent human settlements. Slightly larger, and possessing arable soil and freshwater, are the Marshall Islands. Over population on these islands was of constant concern. Control was obtained through emigration and infanticide. Here marine resources were a necessary part of the otherwise limited vegetable diets. The Gilberts provide a richer and more luxuriant ecosystem. Larger, raised-coral islands support bigger permanent communities. Volcanic islands such as the Hawaiian, Society, and Samoan Islands have many environments, varied plant life, and support complex societies. Largest of the islands are New Guinea and the islands in the Melanesian archipelago with high mountains, rich resources, and isolated and diverse native communities.

Care should be exercised with summaries of generalized fishing patterns. The fisher-horticulturalists of Oceania and South America each developed cultural patterns adapted to their localized ecosystems.

Fishing peoples of the Mariana's assigned the tasks of clearing, planting, and harvesting to men, while women fished the fringing reefs with small hand nets. Portions of their catch were stored in fish ponds along the shore built from stones.[25]

To the southwest, Palau men fished with spears and nets inside the fringing reefs. Women cultivated the taro swamps using digging sticks. Men tended such tree crops as breadfruit, coconuts, and betel nut.

On the larger islands, such as Truk, the fishing technology is more complex and the societies tend to be more stratified. Trukese men fish using traps, weirs, nets, spears, hooks, and lines. They fished bonito in the open ocean. The village chief decides "when communal fishing expeditions occur".[25] Women garden taro and sweet potatoes. "Until recent times the women provided most of the fish consumed, taken from the fringing reefs of the islands. Women did not, however, fish from canoes."[25]

Southeast in the Gilberts, men farm, fish the lagoons and open ocean, and care for artificial fish ponds. Women help the men cultivate, prepare foods, and perform daily household tasks.

The domesticated animals present with fishing and horticulture (see Figure 8) are most often pigs, sometimes chickens. Pigs scavenge and roam freely. They are viewed as an important source of wealth and protein. Most often they are consumed only on ceremonial occasions.

The farmer-fisher peoples of Africa and Oceania (see Figure 8) adopted techniques most suitable to their needs. These include bare hands, hook and line, nets, bow and arrow, spears, traps, weirs, fish ponds, and tethered fish to guide fishermen to their catch. Men usually did the fishing, particularly if it involved travel away from land. Women tended to handle land based subsistence activities.

Women, too, may be responsible for marketing of fish in areas where economic exchange takes place. Among Fanti fishers of Ghana, men traditionally fished and women distributed and processed their catch.[51] With mechanization, women's prominence increased based on their traditional role in financing the fishery. Mechanization increased the cost of equipment to where each fisherman and his kin could not manage the capital requirements. The fishermen then "turned to 'money lenders', most of whom are market women".[51] These women loaned money to buy equipment at interest rates of 50%, and they also got a "share of the catch allocated to the equipment financed".[51]

Figures 7 and 8 imply that fishing complements horticulture. It is easy to see why fishing, more than hunting and collecting, combines well with plant domestication. Farmers clear

FIGURE 9. Trawlers (left) and trollers (right) are among many fishing life styles in a complex society. Jim Larison photo.

land of trees and attempt to control animal and plant pests. The ocean is a good source of animal protein that does not compete with farming. Hunting and collecting are based on nomadism. Farming and fishing nurture fixed settlements. Clark[52] observes for Europe, "With the spread of farming, a noticeable advance may be noted both in off-shore fishing and in the hunting of sea-mammals."

Figures 3, 5, 6, 7, and 8 show cultures adapted to particular ecosystems. Cultures with plant, animal, and aquatic domestication modify their ecosystems more than hunter and gatherers. Their modern counterparts are those whose is built around fishing.

B. Fishing Life Style

Fishing peoples have an intimate and highly integrated ecological-cultural relation. Any change can be expected to affect both their way of life and the natural environment upon which they depend. There is a second type of cultural involvement in fishing. This is the situation where fishing is one among many occupations in complex societies (Figure 9). In this situation, changes in behaviors related to fishing are not likely to change the patterns of integration between the cultural and ecological system. For fishing as an occupational subculture, changes affect the life style of groups of people. In this case the integration of subgroups with the ecosystem is altered.

Groups analagous to the small fishing society are found in the cooperative networks formed among fishermen. These "code groups"[53] consist of several boats who share information on fish locations. These groups of fishermen use secret codes to communicate fishing locations and concentrations.[54,55] Talk of cities may represent the size of the fish concentration. Secret words indicate the species and location. "Mabel wants to go to an anniversary," could mean that sea bass (Mabel) are off Point Conception (anniversary).[53] Like most successful hunter-gatherers, the "tracking" ability of fishermen is a key to success. Successful fishermen are "sensitive not only to what is being said but also to who is talking and the speaker's tone of voice . . . a speaker who is short of breath, for instance, may be working hard in a school of fish".[53]

What behaviors distinguish fishing from other occupations? One difference is in fishermen's willingness to defer personal satisfaction to a later time. This question is investigated by comparing fishermen with millworkers, factory workers, cane workers, and small farmers. Do fishermen, who have to plan ahead to locate fish, select weather conditions, and save income from good fishing for weathering bad times, display more deferred gratification than other workers?

Results of studies in the community of Puerto Real in western Puerto Rico,[56] in Point Judith, Rhode Island,[57] for small-scale fishermen of Panama and Puerto Rico,[58] on Saint Kitts in the Caribbean,[59] and in Arembepe, Brazil[60] suggest that fishermen do have higher levels of deferred gratification. These studies found that the more successful the fishermen, the higher the level of deferred gratification. This is not a universal proposition. Bernard[61] found Greek sponge fishermen are more interested in demonstrating their skill, daring, and superiority over others. Deferred gratification is important from the point of view of development because "injection of development funds into a region with the goal of sustained development through reinvestment of reasonable amounts of profit requires a deferred gratification orientation".[62]

Related to deferred gratification is the risk taking behavior of fishermen. Fishing is a dangerous occupation. When compared with millworkers, Poggie and Gersuny[63] found fishermen were greater risk takers. As risk takers their behavior was highly ritualized to reduce the risk of needless hazards.

Of particular interest is how fishermen organize. When at sea, small groups communicate with one another and offer mutual aid. These support groups have been identified and studied by Acheson and Riedman,[64] Andersen and Wadel,[19] and Stephenson.[55] Their effect is to increase fishing efficiency and safety. Fishermen in these groups share new ideas.

In addition to organizing to help one another, fishermen organize to help the resource. Maine lobstermen,[65] Brazilian swamp fishermen,[66] drift gillnetters, Shoal Harbor, New Jersey fishermen,[67] and many other fishermen form groups to limit effort, protect the resource from overfishing, and improve as well as replenish fishing habitats.

Another characteristic of modern hunters, that is also characteristic of hunter-gatherers, is fishermen's intimacy with the ecosystem from which they derive their livelihood. Each spring men and boats leave Seattle to fish Alaska. They join local residents for the salmon season. Alaska bound salmon fisherman, Upton,[68] summarizes how "That lonely country gets in your blood . . . Some quit for awhile, try other jobs, try staying south, being with their families. But for many it doesn't work . . . .a salmon season in Alaska and all that that means — the trip north to that magnificent place, God's country, the fishing in steep, tree-lined fjords, the glorious days on the beach, and the stormy trip home."

Fishermen comment, "Fishing is a complete experience. It stimulates the mind, body, and spirit." The feelings and attitudes that make people fish are not just the subject of poets, painters, and novelists. Social scientists, too, have shown how love of an activity sometimes outweighs economic gain.[63,69,70]

Farther offshore from the Uptons are fishermen with quite a different life style — the men and women who fish for the distant water fishing nations (Figure 10) of Japan, the Soviet Union, Poland, Bulgaria, Cuba, East Germany, South Korea, and U.S. tuna seiners. Fishing for these nations involves exchanging catches between ecosystems. They bargain with coastal states for fish stocks. These fishermen have a different attachment to the work,

FIGURE 10. Distant water fishing nations practice a different life style. Jim Larison photo.

the trip north, and the glorious days on the beach. They have specific occupational specialities, spend long periods at sea, and fish many ecosystems. Distant water fishermen are more concerned with union agreements and work loads. They do not have the same attachment to the resource.

Distant water fishers move from ground to ground. They search out the most productive areas. Once the best catches are taken, they move off to another area. This leaves the homegrounders like Upton at a disadvantage. He depends on sustained yields available from year to year to support his life style. The distant water fleets have greater mobility, and when the stocks decline, they move on.

A second problem affects the homegrounder. This is the increasing number of people who would like to fish as he does. In many respects the homegrounder faces much the same situation as fishing peoples. Because they lack mobility and depend on a single or limited range of ecosystems, they are susceptible to outside pressures on the local ecosystem. This can be from distant water fishers, resource changes, or changes in markets that reduce demand for the homeground product.

Both fishing peoples, cultures whose way of life depend on one or a few ecosystems and homegrounders, (individuals whose fishing life style depends on a local ecosystem) are endangered by distant water fleets, economic development, and modifications made to ecosystems. That is, people who develop a close and harmonious relation with an ecosystem will be severely disrupted when changes are made to any aspect of this relation.

Upton[68] first came to Southeast Alaska after being a distant water fisherman. He became absorbed in the fishery.

"When I first came to southeast Alaska . . . .it was right on the heels of a year's fishing off northern Chile, the most barren, desolate, and treeless coast that you can imagine . . . And then I started salmon fishing, and it was the best and the most enjoyable fishing that I'd ever done. But the fishing has changed for the worse faster than I would ever have thought it could."[68]

These men remember better times when catches and fish were bigger. Even when new records are set, their concern is maintaining an ecosystem built out of perceptions of the past.

Alaska, like other Pacific Coast provinces and states, has become concerned with Upton's observation. All around the North Pacific Rim experiments in rearing salmon to graze ocean pastures are in progress. The future of salmon concern fishermen, scientists, and politicians. Salmon enhancement is receiving millions of dollars in public and private capital. Will salmon return to productivity levels that Upton first experienced? Yields may return, but domestication carries with it implications for future options. Domestication results in more controlled ecosystems. It simplifies and expands ecological niches. To fishermen, even if productivity is greater; it is not the same quality. It loses its naturalness. A way of life is lost.

C. Aquaculture

The wholeness of existence for fishing peoples such as the Yahgan, where a change alters not just their culture but the whole nature of their cultural and ecological situation, implies a desire for constancy. The concerns of fishermen like Upton over maintaining a life style characterized by freedom, independence, and closeness to nature, too implies maintenance of a way of life. Evolution and acculturation deal with change. One evolutionary change that bears on the cultures of fishing peoples and the life styles of fishermen is aquaculture; the domestication of aquatic plants and animals.

Typically the evolution, resulting in domestication of fish, animals, and plants, leads to more complex societies.[71,72] Domestication requires maintenance of knowledge on breeding experiments. Selection of preferred species is benefited by cooperation so that the best seed stock is developed. Legal structures that protect water and land resources facilitate domestication, but societies invest this effort in only a few fish, plants, and animals.

Once the benefits of domestication have been achieved, human societies extend the range of the plant or animal. Salmon runs have been introduced in New Zealand and Chile, where no runs were known to exist in the Southern Hemisphere. Rice spread throughout South and East Asia, taking over both taro and yam areas.[73] Wheat and barley were introduced into the New World and maize spread to Africa.

Expansion of the ecological niche requires modifications suitable to perfecting the biological adaptation of preferred species. Wild plants and animals occupy these niches and compete with the domestication process. Domestication is more productive and supports complex social systems that are capable of mobilizing to take over culture areas where capture techniques persist. Domestication is an ecosystem expanding technology that sweeps aside fishers, hunters, and gatherers.

But domestication, too, has its limits. During the early 1960s adding domesticated coho salmon to Oregon and Washington rivers and streams raised the quantity of salmon available to fishermen. During the following decade additional coho did not produce any increase in catches. It is difficult to know why boosting the seeding of domesticated coho in the 1970s failed to produce additional returns. One explanation is that some environmental limit, whether biological, chemical, or physical had been reached.

The coho is also a good example of expansion of range. Streams in the Southern Hemisphere, northeast coast of North America, and the Great Lakes all sustain coho runs that did not exist before. Yet coho have not expanded into equatorial waters. The physical limit of temperature restricts this expansion.

Domesticated plants face many of the same ecosystem limits. Productivity gains from the Green Revolution have been achieved. The next level of enhanced terrestrial and aquatic productivity seems to await breakthroughs in applications from knowledge gained about genetic engineering.

D. Philosophy of Life and Aquaculture

Will aquaculture develop with an occidental, monocultural, or the polycultural perspective of oriental culture? Aquaculture reflects the culture of the people who practice it. Trout, tilapia, catfish, and carp farms in the U.S. are run just like the agricultural system. Emphasis is placed on one crop per pond just as farmers till one crop per field. Chinese and East Asians use polyculture. They perceive the environment as consisting of many niches that need to be filled.[13]

This was not always the case. Until 1500 years ago, "Chinese fish culture had been centered around one kind of fish and one kind of practice, carp monoculture".[13] Similar sounding names for common carp and the Emperor led to prohibitions on catching and eating it. Resourceful Chinese sought new species and developed a polycultural system of grass carp who eat submerged aquatic plants, silver carp who consume microscopic plants, bighead carp who feed on microscopic animals, and the mud carp who cleans up the wastes. Pushing further, aquatic polyculture is merged with chicken, duck, and hog rearing, gardening, and silkworms.[13]

Food production technologies are consistent with cultural goals. Indo-European cultures optimize such goals as maximum productivity, economic efficiency, and environmental control with their food producing technologies. Groups focusing on personal freedom, equality, and minimizing disturbance to the environment have proposed small scale, but complex technologies, e.g., the New Alchemists, Amity Foundation, and Ecotope Group. Most studies, which look at the feasibility of aquaculture, focus more narrowly on one or a few technologies. Each of these technologies assume different social system characteristics and needs.[74-79] This may follow the pattern of domestication of terrestial plants and animals where ecosystems are simplified and expand utilizing the techniques that best match organism and cultural characteristics.

E. Future of Aquaculture

What are the likely cattle, pork, and chickens of aquaculture? At the present time, salmon appear to be most like beef. They feed high on the food chain, are costly, and are mainly consumed by those who are well-off economically. Salmon graze over large areas of ocean or the Great Lakes before returning for harvest. The pork analog is carp. These herbivores feed lower on the food chain. They clean up scraps and apparent wastes. Carp have not achieved as high respect in occidental cultures as they have in oriental. The aquatic chicken is tilapia. This fish, of African origin, breeds and grows quickly. Tilapia are advantageous for aquaculture because they are herbivores, and this greatly reduces feeding costs.

All species, except for salmon, are cultured inland. Ocean aquaculture is limited for a number of reasons, but Hanson[77] points to experiments in association with petroleum platforms and the coral reefs of Micronesia as two potentially fruitful places to begin.

Ryther[80] shows that most of the ocean's productivity is associated with upwelling zones. While these are heavily fished areas, they have not been widely used for aquaculture. Marine agriculture or mariculture is most heavily practiced in protected seas, bays, and estuaries where shellfish predominate as the edible fauna. Shellfish, which are equivalent to nuts and acorns, feed low on the food chain. They can be gathered easily and are high in protein.

IV. PATTERN OF FISHING EVOLUTION

The pattern of evolution shows fishing as one part of the subsistence pattern of many

FIGURE 11. Columbia River gillnetter for whom fishing is an occupational life style. Jim Larison photo.

cultures. Precontact cultures along the northwest coast of North America illustrate the most dependence on fishing. Prehistoric data suggest similar levels of dependence for northern Europe, the east coast of North America, and the west coast of South America. Fishing with herding and/or farming characterize Africa, Oceania, and South America.

For fishing peoples, fishing is central to the group's way of life. Fishing as a life style is characteristic of complex societies (see Figure 11). As such it competes with many other occupational subgroups for identity and maintenance of resources to continue the occupational style. The nature of the ecological integration, too, is different. For fishing peoples there is a wholeness between the culture and the ecosystem. For fishing as an occupation, cultural groups form, but the wholeness is more between the individual and the ecosystem.

The Chinese domesticated fish as part of a polycultural system for plant, animal, and aquatic production. Milkfish use oceans as a nursery, and salmon ranching uses the marine environment as a pasture. Shellfish aquaculture takes advantage of the nutrient transport capabilities of marine ecosystems. The pattern of evolution has been to sweep away capture technologies as more productive culture technologies develop. Culturing in occidental cultures is a process of simplification and niche expansion. Oriental aquaculture has evolved during the last 1500 years from monoculture to polyculture.

It seems somewhat paradoxical that human societies, capable of sustaining human life in a broad variety of environments, look toward colonizing space for the future rather than the oceans. Given the dietary habits of humans and the way terrestrial and estuarine ecosystems have been treated, perhaps it is a saving quality that few seek to colonize the oceans where 70% of the earth's oxygen is produced. Defiling the ocean's photosynthesizing plant life could end the process of human evolution. Only a few, however, have tickets for space. Those of us left on the spaceship earth will have to look around for ways to sustain ourselves.

V. ACCULTURATION

Acculturation is a biosphere process that links ecosystems. It is the process of contact between cultures that evolved in varied ecosystems. While human societies have not sought to colonize the oceans, many peoples have sought to link ecosystems and colonize one another. This is one type of acculturation where a dominating culture introduces its ways to another group, often for what are articulated as humanitarian reasons. It usually occurs between cultures at different levels of technological evolution. The reason for contact is exchange between ecosystems. European explorers contacted other cultures and increased our knowledge of the earth's surface. Missionaries and traders followed explorers, trying to introduce religion and win technological converts.

European traders wanted to open up new ecosystems to world markets. Entrance into the world's market economy is not without hazards, particularly for ecosystem cultures that have resources in demand. Mowat tells of an Innuit man (Eskimo word for people) by the name of Kakumee who was introduced to the world's goods by little gifts from a trader. According to Mowat,[81] "the things Kakumee was given, made him sick with desire, yet he was told they were not his to own but were to be traded for furs". Many of Kakumee's people also sought furs to trade. They became dependent on food items bought at the trading post.

World War I changed people's demand for furs. With the fur trade cut off, starvation wiped out many of the people. After the war the market came back. But in the late 1930s, the traders withdrew again. More people starved. According to Mowat[81] Kakumee survived because, "he drove his dog team to the camp of some fortunate man who had made a lucky hunt and took what he wanted". Kakumee then summarized the condition of his people, "You," he said, "are rich. You are very rich, white man! Richer in tea, and in rifles and shells than we of the People. And yet we too are rich! Rich in graves and in ghosts — and this is your doing."[81]

Some say Mowat overdramatized the problem. Certainly he did not get the policy changes needed to help Kakumee's people. Most of Mowat's readers are sympathetic but skeptical. There must be some flaw in Kakumee or Innuit culture, they say.

A more scientific study of acculturation is Firth's observations of changes in the Malay fishing village of Kelantan. Firth[82] did field work among the Kelantan fishermen in 1939 to 40, 1947, and 1963. During each of these time periods, he observed change from a relatively unmechanized lift-net and gill-net to a more mechanized purse-seine fishery as the villagers were encouraged to become part of an expanding market system.

Mechanization resulted in fewer people having a share of the capital resources needed for fishing. Ownership of nets and boats declined between 1940 and 1963. Those without these capital resources increased from one-third to nearly two-thirds of the fishermen with no interest in either net or boat.

Firth also provides data on change in the shares for lift-net earnings in 1940 and seine earnings in 1963. These, too, indicate loss of income that would provide the capital to pursue the fishery. The average daily wage in 1963 actually declined from 1940 by one-third. Firth[82] summarizes his results.

"There is little doubt that during the last thirty years the position of the peasant in Kelantan has tended to change, particularly in the direction of greater differentiation in levels of wealth . . . . Much greater returns to fishing in modern conditions, accompanied by or resulting from much greater capitalization, has resulted in a marked drop in the percentage of earnings going to labour."[82]

The same pattern is observed in northeastern Brazil. Here the mangrove swamps provide refuge for returnees from Amazon Basin development.[66] Development of the Amazon was seen by the Brazilian government as a way of alleviating poverty in northeastern Brazil where peasants were flocking, forced off their lands by earlier sugar, rubber, and coffee plantation development. The Amazon soils proved too poor and the forests too thick. Many peasants were unable to get jobs in the northeast, and denied other opportunities, became swamp fishermen.

But entrepreneurs, "as hungry to open up new markets as the land-starved peasantry was to sell its labor, introduced nylon nets".[66] FAO tests showed nylon nets increase catches. What FAO and the development experts did not study was the cultural system of the swamp fishermen. The swamp fishermen were adapted to the local ecology. They evolved mechanisms for distributing the catch, cooperative fishing etiquette, mediating disputes, and sharing and conserving resources.

As is often the case, the swamp fishermen rejected the nylon net innovation. There was a brief period of increased catches, but within a decade catches began to decline. Fewer net crews were needed. They did not oppose the more effective technology, but the technology did not fit their culture. Most could not afford the loan payments. Thus, the nets became the property of wealthy middlemen and, like the Malay fishermen, the peasants became laborers. The system of territorial claims broke down.

Some do benefit from economic development. Ward[83] gives the example of two Chinese families in the village of Cheung Chau. Lau Kam-Ch'ing owned two wind driven deep-sea trawlers in 1950. He had a series of good seasons and was able to obtain a loan to mechanize. Lau Kam-Ch'ing did not fish, his family thought he brought bad luck. He was the shore-based manager for the family's trawling business. His brothers operated the fishing vessels. The loan for mechanizing enabled them to fish new grounds and the family prospered. The family purchased a third trawler and began loaning money to other fishermen. Lau Kam-Ch'ing's son was the first from the community to attend fisheries college in Japan.

Chung Fuk-Shun, too invested in a new boat. His son, other family members, and two hired men operated the boat. Chung Fuk-Shun's timing was not as good as that of Lau Kam-Ch'ing. He found he was on the treadmill of overcapitalization. In 1963, two years after mechanizing, Chung Fuk-Shun bought an even more powerful engine. He hired four more men. His fuel bills doubled, but he was only catching the same amount as he had before. In addition to greater mechanization, "No fisherman now fails to see the need for education. Coxswains' engineers' certificates, successful marketing, coping with government regulations, all depend on it".[83]

It was to no avail. Chung's daughter, who had fished for him, married a nonfisherman. He sold his boat in 1966 and sent his son to work in England for a friend who owned a Chinese restaurant.

U.S. development efforts are technologically rich and primarily oriented to exchanging goods between ecosystems. Critchfield[84] relates how one U.S. development effort affected local fishermen in Mauritius. The 600 fishermen of Grand Gaube had sought government aid in procuring boats large enough to get outside the inshore reefs where they usually fished. The resource was being depleted as the fishermen sought to meet the food needs of the rapidly expanding island population.

One day an American fisheries expert, sent by the United Nations, showed up in the village. The American agreed that it was important to get the fishermen off the depleted inshore reefs, but he said that the fishermen would have to work through the government. He also encouraged the fishermen to form an association. After pledging to return in about two weeks, the fisheries expert left Grand Gaube.

Two months later the fishermen had heard nothing. They hired a taxi and drove to the American's office. He had worked on the problem and found that an association already existed. The fishermen explained that few belonged to the existing organization because it had been established to get fishermen to report anyone using illegal dynamite or mosquito nets while fishing at night.

The fisheries expert, too, had secured the help of the American Embassy. They were willing to purchase materials if the fishermen would contribute their labor to build boats. In Grand Gaube, however, marine carpenters were a specialized occupation. They would not contribute their labor for the benefit of fishermen. The fisheries expert said the project had to be one of "self-help". The American said, "There are too many fishermen. There are not enough fish in the lagoon . . . . They have got to find something else to do." Both the fishermen and the American were becoming irritated with the remorselessness of the situation. A big car drove up and a woman mouthed some words to the fisheries expert. He rose and said, "I'm sorry, I have another engagement." He shook the fishermen's hands and left. The fishermen left feeling the obstructions of regulation, understanding, and culture that formed the barrier between them.

Little investment is made studying the cultural and ecological systems into which new technologies are placed. Usually when investment is made in studying the cultural system, it is after some problem has arisen.

Such a situation occurred in Surinam.[85] In 1956 a village containing 35 homes with bath, kitchen, two bedrooms, living room, electricity, running water, and screening was built by the government for fishermen. Financing and planning was provided by the U.S. Three years later only nine homes were occupied. An anthropologist was asked to determine why fishermen did not choose to live in these fine, new homes.

Oberg and Van Dijk[85] found that the fishermen needed at least 50 yards of river frontage for drying, repairing, and tarring nets; for boat storage; and for a garden, chicken house, and space for hired hands. Fishermen felt the need to be near their boats and gear to prevent theft or damage. The closely packed village away from the river did not meet these needs and was abandoned.

For fishing peoples the twin processes of evolution and acculturation press their extinction. The hunter-fisher-gatherer way of life where small groups migrate to capture their food supply does not fit the evolution to cultures where sedentary life styles, large populations, and highly managed ecosystems are the way of life. The trend of evolution, if it is to continue, offers little future for fishing peoples.

Acculturation poses no less pessimistic a future. Here the biosphere explorers, traders, and scientists seek to interchange technologies, values, and life styles. Most economic development situations see the life styles of fishing peoples as technologically and economically inefficient. The solution, therefore, is to modify their ties with the ecosystem and provide more efficiency through economic development. The problem with this thinking is that cultures and ecosystems are integrated wholes. Fishing people's culture is not quaint and unique; it represents their adaptation to an ecosystem. Make one change and the whole system of interrelationships changes.

While fishing life styles are occupational subcultures of a larger cultural system, they too are unlikely to survive the pressures of domestication and competition for attention. The old ways pass as fishermen adjust to more control that comes with overexploitation and managing ecosystems for greater productivity through domestication.

Economic development is assumed to improve the overall well-being of those with less developed technologies. The result in most cases is that a few are better off; the majority are not (Figure 12). The examples from Cordell,[66] Critchfield,[84] Firth,[82] Mowat,[81] Oberg and Van Dijk,[85] and Ward[83] show this. For other cases see Alexander,[86] Berleant-Schiller,[87] Evans,[88] Keddie,[90] Landberg and Weaver,[91] Nietschmann,[92] Pollnac and Suitinen,[93] Price,[94] and Spoehr.[95]

But economic development is a process that serves mixed goals. While it causes greater inequality, not to make available "advanced" technology also withholds the opportunity to pursue the material life style which the world's biosphere traders have achieved. Thus, while the ecological risks are great when cultures enter the world market system with only a single

FIGURE 12. In most cases of economic development a few are better off; the majority are not. Jim Larison photo.

primary resource, so too, are the political risks of trying to select who should or should not have the opportunity to adopt new technologies.

What new innovations, then, might we look for? When humans 5000 years ago ushered in transoceanic travel, they ushered in linking ecosystems through trade. When 3000 years ago they invented aquaculture, they heralded more controlled ecosystem adaptations that are to the disadvantage of fishing peoples. What new technologies may affect how we contemplate and use the oceans?

VI. COMMUNITIES AT SEA

Economic development shows that the process of trying to change old systems with new approaches seldom works. Several factors explain this. One is that over long periods of time cultures and ecosystems have become ecological wholes, and the human behaviors fit the available options. Economic development that does not recognize the interdependence between cultures and their ecologies usually fails. Second, acculturation which brings new technologies developed in other systems assumes what worked for the dominant culture will work for the developing culture. Time, space, and ecology are different and this assumption does not hold. Finally, the peoples linking ecological systems so as to exchange goods throughout the biosphere seldom have the broad picture. They are mainly trying to enhance their own ecological situation.

It may be easier to start fresh in creating the new cultural-ecological relation. This implies that it is easier to break new ground than to recultivate old. The missionaries of change will continue to ply their trade. They will have some success. There also appear to be new opportunities that can utilize the current state of cultural evolution and integrate knowledge resulting from the contact between cultures.

Economic activity related to fisheries holds the potential for stimulating future ocean habitation. The creation of 360 km economic zones to protect fish resources is an institutional change, that may foster new uses of the oceans as terrestrial oriented humans expand into this new ecosystem.

If we project the future impacts of the Law of the Sea Conference in extending countries' economic zones to 360 km, the limitations on distant water fishing comes to mind first. This may be a transitory impact as the order of the world's fishing nations has not changed dramatically. The large fishing nations have worked out agreements with many coastal states.

As coastal states realize the stabilizing effect of the 360 km economic zone, this should enhance the potential of establishing offshore communities. The needs in transshipping energy cargo, generating energy from thermal gradients, food production in the nutrient rich upwelling zones, and recreation opportunities all point to the utility of developing offshore communities in the upwelling zones.

Only excessively crowded cities like Hong Kong, Bangkok, Kobe, Calcutta, Canton, Singapore, Shanghai, and Rangoon actually use their water areas for habitat. More often human communities fill in estuaries to create more land close to shore. Architects and planners have failed to see the nearshore ocean ecosystem as a place for community development. Sixty percent of the world's population lives within 100 km of the coastline. Community development in the ocean is a natural extension of people's move toward the sea. The 360 km economic zone should provide the political stability to make such marine colonization possible.

Two factors hold this colonization back. One is the human fear and awe of the ocean's power. We have seen our most invincible structures crumble before the ocean's awesome power. Second, while we have perfected exchanging resources between ecosystems, we have not developed enough appreciation for the biosystem as a higher order of ecological integration. Our waste disposal practices are too filthy to safely inhabit the oceans. Human society cannot safely inhabit the oceans until recycling, reuse, and the patterns of ecosystem people working in harmony with nature become a part of the pattern of human interaction in the biosystem. The risk is too great. The oceans are our primary oxygen producer. Humanity cannot take terrestrial habits of waste disposal into the fluid medium of the oceans. There is too much movement and mixing, too much at stake. We cannot treat the oceans like the moon and leave our garbage for some future generation.

It is probably fortunate that humans have not colonized the nearshore ecosystem. How long will this oversight continue? It will not be long, if the nearshore ecosystem finds a place in the grand schemes of biosphere peoples. Before that happens, hopefully those of an ecosystem persuasion will develop some of the perspective characteristics of biosphere peoples. One would also hope that biosphere peoples develop greater appreciation for the linkages between the many ecosystems they seek to exploit and modify. Stepping into space to look down on the planet Earth may be a first step in expanded use of the nearshore ecosystem.

ACKNOWLEDGMENT

We appreciate the comments of T. C. Hogg and S. B. Peterson on this manuscript. This chapter results from research sponsored, in part, by the Oregon State University Sea Grant College Program, supported by NOAA Office of Sea Grant, Department of Commerce, under Grant 048-MO1-44.

REFERENCES

1. **Dasmann, R. E.,** *Wildlife Biology,* 2nd ed., John Wiley & Sons, New York, 1981.
2. **Gemill, M.,** *The Dispossessed,* film, Ballis, G., producer, 1970.
3. **Slickpoo, A. P. and Walker, D. E., Jr.,** *Noon Nee-Me-Poo, Culture History of the Nez Perces,* Nez Perce Tribe of Idaho, Lapwai, 1973.
4. **Lenski, G. and Lenski, J.,** *Human Societies, an Introduction to Macrosociology,* McGraw-Hill, New York, 1978.
5. **Heyerdahl, T.,** *Early Man and the Ocean: a Search for the Beginnings of Navigation and Seaborne Civilizations,* Doubleday, Garden City, N.Y., 1979.
6. **Leithauser, J. G.,** *Worlds Beyond the Horizon,* translated by Merrick, H., Knopf, New York, 1955.
7. **Suggs, R. C.,** *The Island Civilizations of Polynesia,* New American Library, New York, 1960.
8. **de Bisschop, E.,** *Tahiti Nui,* translated by Edward Young, McDowell, Obolensky, Inc., New York, 1959.
9. **Sauer, C. O.,** Seashore-primitive home of man?, *Proc. Am. Philosoph. Soc.,* 106, 41, 1962.
10. **Baird, S.,** Reports of special conferences with American Fish-Culturists' Association and state commissioners of fisheries, Report of the Commissioner of Fish and Fisheries for 1872-73, Washington, D.C., 1874, 758.
11. **Bullen, F. T.,** *The Log of a Sea-Waif, Being Recollections of the First Four Years of My Sea Life,* Smith, Elder & Co., London, 1901.
12. **Deacon, G. E. R.,** *Seas, Maps, and Men,* Crescent Books, London, 1962.
13. **Ling, S.,** *Aquaculture in Southeast Asia, A Historical Overview,* University of Washington Sea Grant Publications, Seattle, Wash., 1977.
14. **Barry, H.,** Note to readers, *Ethnology,* 6, 489, 1967.
15. **Naroll, R.,** The Culture-bearing unit in cross-cultural surveys, in *A Handbook of Method in Cultural Anthropology,* Naroll, R. and Cohen, R., Eds., Columbia University Press, New York, 1970, 721.
16. **Chaney, R. P.,** Scientific inquiry and models of socio-cultural data patterning: an epilogue, in *Models in Archaeology,* Clarke, D. L., Ed., Methuen & Co. Ltd., London, 1972, chap. 26.
17. **Acheson, J. M.,** Anthropology of fishing, *Ann. Rev. Anthropol.,* 10, 275, 1981.
18. **Andersen, R. and Stiles, G.,** *Man and the Sea: Human Factors in Marine Environments,* Simon and Schuster, New York, 1971.
19. **Andersen, R. and Wadel, C.,** North Atlantic Fishermen, Memorial University of Newfoundland, Institute of Social and Economic Research, Newfoundland Social and Economic Papers No. 5, 1972.
20. **Casteel, R. W. and Quimby, G. I.,** *Maritime Adaptations of the Pacific,* Mouton Publishers, Paris, 1975.
21. **Coon, C. S.,** *The Hunting Peoples,* Little, Brown and Company, Boston, 1971, chap. 6.
22. **Hasslof, O., Henningsen, H., and Christensen, A. M., Jr.,** *Ships and Shipyards, Sailors and Fishermen, Introduction to Maritime Ethnology,* Rosenkilde and Bazzer, Copenhagen, 1972.
23. **Pollnac, R. B.,** Continuity and Change in Marine Fishing Communities, International Center for Marine Resource Development, University of Rhode Island, 1976.
24. **Smith, M. E.,** *Those Who Live From the Sea,* West Publishing Company, St. Paul, 1977.
25. **Alkire, W. H.,** An introduction to the peoples and cultures of Micronesia, *Module in Anthropology,* No. 18, Addison-Wesley, New York, 1972.
26. **Andersen, R.,** *North Atlantic Maritime Cultures,* Mouton, The Hague, 1980.
27. **Drucker, P.,** *Indians of the Northwest Coast,* The Natural History Press, Garden City, N.Y., 1955.
28. **Harding, T. G. and Wallace, B. J.,** *Cultures of the Pacific,* The Free Press, New York, 1970.
29. **Landberg, L. C. W.,** *A Bibliography for the Anthropological Study of Fishing Industries and Maritime Communities,* University of Rhode Island, Kingston, 1973.
30. **Murdock, G. P.,** African Cultural Summaries, 9 Volumes, HRAF Press, New Haven, Conn., 1958.
31. **Murdock, G. P. and O'Leary, T.,** *Ethnographic Bibliography of North America,* Human Relations Area Files Press, New Haven, 1975.
32. **Oliver, D. L.,** *The Pacific Islands,* Doubleday, Garden City, N.Y., 1961.
33. **Poggie, J. J., Jr., Stuster, J., Pollnac, R. B., Carmo, F., McCay, B. J., McGoodwin, J. R., Orbach, M. K., and Petterson, J. S.,** Maritime anthropology, *Anthropol. Q.,* 53, 1, 1980.
34. **Steward, J. H.,** *Handbook of South American Indians,* Vol. 1-7, Bureau of American Ethnology Bulletin 143, 1946.
35. **Vayda, A. P.,** *Peoples and Cultures of the Pacific,* The Natural History Press, Garden City, N.Y., 1968.
36. **Hewes, G. W.,** Aboriginal Use of Fishery Resources in Northwestern North America, Ph.D. Dissertation, Department of Anthropology, University of California, Berkeley, 1947.
37. **Suttles, W.,** Northwest coast, *Handbook of North American Indians,* Vol. 7, Smithsonian Institution, Washington, D.C., in press, 1982.
38. **Boas, R.,** *The Religion of the Kwakiutl Indians,* Columbia University Press, 1930, 288.
39. **Mead, M.,** *Growing Up in New Guinea,* William Morrow, New York, 1930.

40. **Mead, M.,** *New Lives for Old Cultural Transformation — Manus 1928-1953,* William Morrow, New York, 1956.
41. **Fortune, R. F.,** *Manus Religion,* American Philosophical Society, Philadelphia, 1935.
42. **Cooper, J. M.,** The Yahgan, Bureau of American Ethnology Bulletin 143, 1946.
43. **Walker, D. E.,** Mutual Cross-Utilization of Economic Resources in the Plateau: An Example from Aboriginal Nez Perce Fishing Practices, Washington State University, Laboratory of Anthropology Report of Investigations No. 41, 1967.
44. **Steward, J. H.,** Basin-Plateau Aboriginal Sociopolitical Groups, Bureau of American Ethnology Bulletin 120, 1938.
45. **Thomas, D. H.,** A computer simulation model of Great Basin Shoshonean subsistence and settlement patterns, in *Models in Archaeology,* Clarke, D. L., Ed., Methuen & Co. Ltd., 1972, chap. 17.
46. **Snow, D. R.,** *The Archaeology of New England,* Academic Press, New York, 1980.
47. **True, D. L.,** Early maritime cultural orientations in prehistoric Chile, in *Maritime Adaptations of the Pacific,* Casteel, R. W. and Quimby, G. I., Eds., Mouton, The Hague, 1975, chap. 4.
48. **Moseley, M. E.,** *The Maritime Foundations of Andean Civilization,* Cummings Publishing Company, Menlo Park, N.J., 1975.
49. **Stack, B. L. and Voorhies, B.,** *Prehistoric Coastal Adaptations. The Economy and Ecology of Maritime Middle America,* Academic Press, New York, 1978.
50. **Siskind, J.,** *To Hunt in the Morning,* Oxford University Press, New York, 1973.
51. **Christensen, J. B.,** Motor power and women power: technological and economic change among the Fanti fishermen of Ghana, in *Those Who Live From the Sea,* Smith, M. E., Ed., West, New York, 1977, 71.
52. **Clark, J. G. D.,** *Prehistoric Europe. The Economic Basis,* Philosophical Library, New York, 1952, chap. 3.
53. **Stuster, J.,** Where "Mabel" may mean "sea bass", *Nat. Hist.,* 87, 64, 1978.
54. **Andersen, R.,** Those fisherman lies: custom and competition in North Atlantic fisherman communication, *Ethnos,* 1, 154, 1973.
55. **Stephenson, G. O.,** Pushing for the Highline: The Diffusion of Innovations in the Oregon Otter Trawl Fishery, M.A.I.S. Thesis, Oregon State University, Corvallis, 1980.
56. **Poggie, J. J., Jr.,** Deferred gratification as an adaptive characteristic for small-scale fishermen, *Ethnos,* 6, 114, 1978.
57. **Pollnac, R. B., Gersuny, C., and Poggie, J. J., Jr.,** Economic gratification patterns of fishermen and millworkers in New England, *Human Organ.,* 34, 1, 1975.
58. **Pollnac, R. B. and Poggie, J. J., Jr.,** Economic gratification orientations among small-scale fishermen in Panama and Puerto Rico, *Human Organ.,* 37, 355, 1978.
59. **Aronoff, J.,** *Psychological Needs and Cultural Systems,* Van Nostrand and Reinhold Company, 1967.
60. **Kottak, C. P.,** The Structure of Equality in a Brazilian Fishing Community, Ph.D. Dissertation, Columbia University, New York, 1966.
61. **Bernard, H. R.,** Kalymnos, Island of the sponge fisherman, in *Technology and Social Change,* Bernard, H. R. and Pelto, P. J., Eds., The Macmillan Company, New York, 1972, 277.
62. **Pollnac, R. B.,** Panamanian Small-Scale Fishermen: Society, Culture, and Change, University of Rhode Island International Center for Marine Resources Development, Marine Technical Report No. 44, Kingston, 1977.
63. **Poggie, J. J., Jr. and Gersuny, C.,** Fishermen of Galilee, University of Rhode Island Marine Bulletin Series No. 17, 1974.
64. **Acheson, J. M. and Riedman, R.,** Technological innovation in the New England fishing industry: an examination of the Downs and Mohr hypothesis, *Am. Ethnol.,* 9, 20, 1982.
65. **Acheson, J. M.,** Territories of lobstermen, *Nat. Hist.,* 81, 53, 1972.
66. **Cordell, J.,** Swamp dwellers of Bahia, *Nat. Hist.,* 87, 62, 1978.
67. **McCay, B. J.,** Optimal foragers or political actors? Ecological analyses of a New Jersey fishery, *Am. Ethnol.,* 8(2), 356, 1981.
68. **Upton, J.,** *Alaska Blues,* Ricwalt Publishing Co., Seattle, Wash., 1977.
69. **Smith, C. L.,** Fishing success in a regulated commons, *Ocean Devel. Int. Law,* 1, 369, 1974.
70. **Smith, C. L.,** Satisfaction bonus from salmon fishing: implications for economic evaluation, *Land Econ.,* 57, 1981.
71. **Service, E. R.,** *Origins of the State and Civilization,* W. W. Norton, New York, 1975.
72. **Wittfogel, K. A.,** *Oriental Despotism,* Vintage Books, New York, 1981.
73. **Barrau, J.,** L'Humide et le sec: an essay on ethnobiological adapatation to contrastive environments in the Indo-Pacific area, in *Peoples and Cultures of the Indo-Pacific,* Vayda, A. P., Ed., The Natural History Press, Garden City, N.Y., 1968.
74. **Bardach, J. E., Ryther, J. H., and McLarney, W. O.,** *Aquaculture: The Farming and Husbandry of Freshwater and Marine Organisms,* John Wiley & Sons, New York, 1972.

75. **Borgese, E. M.,** *Seafarm, The Story of Aquaculture,* Harry N. Abrams, Inc., New York, 1980.
76. **Brown, E. E.,** *World Fish Farming: Cultivation and Economics,* AVI Publishing Company, Westport, Conn., 1977.
77. **Hanson, J. A.,** *Open Sea Mariculture, Perspectives, Problems, and Prospects,* Dowden, Hutchinson, and Ross, Inc., Stroudsburg, 1974.
78. **Hickling, C. F.,** *Fish Culture,* Faber and Faber, London, 1971.
79. **Pillay, T. V. R.,** *Planning Aquaculture Development, An Introductory Guide,* Fishing News Books, Farnham, 1977.
80. **Ryther, J. H.,** Photosynthesis and fish production in the sea, *Science,* 166, 72, 1969.
81. **Mowat, F.,** *People of the Deer,* Pyramid, New York, 1951.
82. **Firth, R. A.,** *Malay Fishermen: Their Peasant Economy,* 2nd ed., Revised, Routledge & Kegan Paul, London, 1966.
83. **Ward, B. E.,** Chinese fishermen in Hong Kong: their post-peasant economy, in *Social Organization, Essays Presented to Raymond Firth,* Freedman, M., Ed., Aldine, Chicago, 1967, 271.
84. **Critchfield, R.,** *The Golden Bowl Be Broken, Peasant Life in Four Cultures,* Indiana University Press, Bloomington, 1973, chap. 2.
85. **Oberg, K. and Van Dijk, F.,** The Fishermen of Surinam, Surinam-American Technical Cooperative Service, Parimuribo, Surinam, 1960.
86. **Alexander, P.,** Innovation in a cultural vacuum: the mechanization of Sri Lanka fisheries, *Human Organ.,* 34, 333, 1975.
87. **Berleant-Schiller, R.,** Development proposals and small-scale fishing in the Caribbean, *Human Organ.,* 40, 221, 1981.
88. **Evans, E. B.,** Marine scientific and technological assistance to developing countries: science for development or technology for malintegrated growth, *Econ. Devel. Cultural Change,* 24, 375, 1976.
89. **Forman, S.,** *The Raft Fishermen,* Indiana University Press, Bloomington, Ind., 1970.
90. **Keddie, W. H.,** Fish and futility in Iranian development, *J. Devel. Areas,* 6, 19, 1971.
91. **Landberg, L. C. W. and Weaver, P. L.,** Maendeleo: economic modernization in a coastal community of northeastern Tanzania, in *Rethinking Modernization,* Poggie, J. J., Jr. and Lynch, R. N., Eds., Greenwood Press, Westport, Conn., 1974, chap. 8.
92. **Nietschmann, B.,** Hunting and fishing focus among the Miskito Indians, Eastern Nicaragua, *Human Ecol.,* 1, 41, 1972.
93. **Pollnac, R. B. and Sutinen, J. G.,** Small Scale Fisheries in Central America: Acquiring Information for Decision Making, University of Rhode Island International Center for Marine Resource Development, Kingston.
94. **Price, R.,** Caribbean fishing and fishermen: a historical sketch, *Am. Anthro.,* 68, 1363, 1966.
95. **Spoehr, A.,** *Protein from the Sea: Technological Change in Philippine Capture Fisheries,* University of Pittsburgh, Ethnology Monographs, No. 3, Pittsburgh, 1980.

Chapter 3

SHIPPING IN THE GREAT LAKES

Harry C. Brockel, Eric Schenker, and Harold M. Mayer

TABLE OF CONTENTS

I. EDITOR'S INTRODUCTION

The Great Lakes together constitute the largest body of navigable freshwater in the world. With a relatively small drainage basin of 295,000 mi^2, the Lakes themselves have a surface area of just under 100,000 mi^2. Elevations above sea level range from an average of 248 ft for Lake Ontario to 602 ft for Lake Superior. From the head of the Lakes at Duluth and Chicago, distances to the head of tidewater at Montreal are 1,337 and 1,244 mi, respectively, and Montreal is 800 mi from the Atlantic Ocean. While a significant proportion of the Great Lakes-St. Lawrence route involves open-water navigation, the connecting channels between the Lakes, and much of the St. Lawrence River, especially above Montreal, involve transit through confined channels. These place serious limitations upon the dimensions, especially the drafts, of vessels. Also the severe ice of northern winters imposes further seasonal hazards and constraints. Despite these limitations the Great Lakes have developed into a major avenue of commerce.

Spanning the entire industrial core of the U.S. and providing a marine avenue to the sea, the Great Lakes form the basis of a unique transportation system. Economically and culturally the lakes have played a major role in the growth and development of commerce in this nation. Fifty years ago the Great Lakes fleet pioneered automated bulk cargo handling with the first self-unloading ships. In 1959, the completion of the St. Lawrence Seaway signaled a major increase in Great Lakes shipping with upwards of 15 million tons/year moving directly overseas from Great Lakes ports. In the early 1970s the successful application of ice breaking and ice control techniques has demonstrated the feasibility of extending the Great Lakes shipping season from just under 8 months to a full 12 months.

Even though this strategically located and dynamic system has sometimes led the way in technological innovation, other technological changes, such as containerization and larger ocean vessels, are conspiring to reduce the volume of ocean shipping to Great Lakes ports. Despite these and other trends, which could adversely affect the Great Lakes-St. Lawrence Seaway Transportation System, the system, nevertheless, remains a vital link in the industry-resource complex upon which the U.S. depends.

The early development of the Great Lakes transportation system is a saga of the opening of the interior of the North American continent by European explorers and traders by water travel. As trading centers became cities and the rich resources of the region were tapped and exploited, the need for transport of goods and people increased. Predating the railroads, water transportation was the key to opening up and settling the heartland of North America. Great Lakes sailing vessels and later powered ships evolved as a uniquely adapted system, namely the package freighters.

Harry Brockel tells this story with a personal and sensitive understanding of the human challenges and the cultural changes which shaped the system. His account ranges from the Indian canoe and the bateau of the early French traders to the opening of the St. Lawrence Seaway and the Great Lakes shipping revolution of the 1960s and 1970s.

II. PACKAGE FREIGHTERS OF THE GREAT LAKES — SAIL AND STEAM

A. The Trade and the Ships

"Package freight" on the Great Lakes would be general cargo on the oceans. The term applied to cargo carried by water, usually of high or medium value, handled in boxes, barrels, bales, cases or as an individual unit, as distinguished from bulk cargoes.

The original package cargo carrier on these waters was of course the Indian canoe which, with the arrival of the white man, became a transporter of furs in exchange for packets of trade goods. The canoe could be of various sizes up to 30 ft and for various numbers of paddlers. It was followed by the bateau of the French fur traders who, by canoe and by

FIGURE 1. The Griffon, built in the late 1670s, was the first vessel to sail the Great Lakes west of Niagara. Picture from a painting made at Green Bay by Cuthbertson. (Courtesy of the State Historical Society of Wisconsin.)

bateau, explored and colonized the entire Great Lakes and their connecting rivers and waterways, and then used these craft to establish their flourishing fur trade throughout the Great Lakes, up and down the Mississippi, and well into western Canada.

B. The Age of Sail

Robert Cavelier, Sieur de La Salle, built the *Griffon* at Niagara, the first vessel to sail the Great Lakes west of Niagara. According to Barry[1] she was a small galleon, a Holland type of vessel, probably about 70 ft long, three-masted, with a crew of about six men (Figure 1). She sailed west from Niagara on August 6, 1679, bound for Detroit, Mackinac, and Lake Michigan. At Washington Island, near Death's Door Passage, an advanced group of La Salle's men had assembled a quantity of furs. Pressed for funds, La Salle decided to attempt a passage back to Niagara, whence the furs could be transshipped to Montreal for sale. On September 18, 1679, the Griffon sailed from Washington Island. Within hours, a violent storm of 4-day duration swept the upper Lakes, and the Griffon disappeared, somewhere in upper Lake Michigan, the Straits of Mackinac, or Lake Huron. Thus ended mysteriously the short career of the first package carrier on the upper Lakes, or in Lake Michigan.

The bateau and the canoe continued as the watercraft of the region for trade and for war, until the events of the Seven Years' War (1757 to 1763) brought about a naval competition between the French in Canada and the British in their American colonies. Dockyards were established on both sides of the border, and small fleets of schooners and sloops were launched. The first naval engagement occurred in June 1756 on Lake Ontario between three British craft and four larger and better-armed French sailing vessels.

A considerable naval rivalry developed between England and France during the French and Indian War, and between England and America during the Revolutionary War. A goodly number of armed sailing craft were built in these struggles for control of the freshwater seas.

With the treaty of Paris of 1783, peace returned, and boundaries were established between British Canada and the new American nation. The Northwest Ordinance of 1787 was a far-seeing declaration by the new congress as to navigation and commerce in these waters. Between 1785 and 1789, the first mercantile vessels since the Griffon appeared, built variously at Sault Ste. Marie, Detroit and Kingston, Ontario. An 80-ton sloop, Detroit, launched near Detroit in May 1793, was the first merchant vessel on the upper Great Lakes to fly the flag of the U.S.

The age of mercantile sail on the Great Lakes extended from the 1780s until 1938, when the final voyage under sail was made by the J. T. Wing, a three-masted schooner in the lumber and pulpwood trade. In September 1930, the schooner Our Son was lost, in northern Lake Michigan, en route to Muskegon with a cargo of pulpwood. Our Son, a 720-gross ton schooner, might be selected as typical of the several thousand wooden sailing vessels which plied these waters and were the package freighters for the region until displaced entirely by steam and iron. Built in 1875 at Black River, Ohio, she survived as a freshwater work horse for 55 years, and must have fully earned her keep. She was 182 ft long, of 35-ft beam, and could carry 1,000 tons of ore, or an impressive 40,000 bushels of grain. In the lumber trade, these ships would fill their holds with lumber or cedar posts or tan bark, and carry prodigious deck loads of forest products which left only the space essential to work the ship.

In a region where lumber was a plentiful raw material, and where skilled craftsmen usually were in good supply, the building of wooden sail ships flourished as a regional art, whether native or immigrant. There was no formal documentation of sailing craft for many decades, and every sheltered cove or small marine settlement had shipbuilding potential. No census of lake sailing vessels seems to exist, but, from the sources available, it would not be an exaggeration to state that between the Revolutionary War and World War I, some thousands of wood-hulled sailing vessels, primarily sloops and schooners, were launched in the Great Lakes.

Although accurate ship building records were not maintained, the number of vessels lost gives some idea of the number sailing the Lakes. Wright[2] has documented 1,324 wrecks in the coastal waters of Lake Michigan alone. Of these, 700, or more than half, were of the age of sail. Heden states, "The records of the United States Commissioner of Navigation from 1878 to 1898 list 5,999 shipwrecked vessels on the Great Lakes of which 1,093 were total losses."[3]

Stonehouse reports 500 recorded shipwrecks in Lake Superior.[4] Unofficial estimates are that 6,000 shipwrecks have occurred in the five Great Lakes since the advent of the white man. This is not intended as a dissertation on shipwrecks, and is a dissertation on history only to the extent necessary to make the point that the waters of the Great Lakes teemed with great numbers of sailing vessels. These were the original lake package freighters, serving every port, large or small, and carrying every conceivable aspect of the regional commerce.

Between Washington Island, Detroit Island, and the Lake Michigan shore of Door County is a natural passage between Green Bay's water and the main lake. A funnel for swift currents and high winds and cluttered with treacherous, rocky islands, this passage soon became known for its deadly nature by the early French settlers. They named it "Porte des Morts" Passage, anglicized as "Death's Door", and from which Door County drew its name.

A lighthouse keeper of literary inclination was Martin N. Knudreu, keeper of the light on Pilot Island, in Death's Door Passage. A diary he kept records that from 1872 to 1889, shipwrecks in the passage averaged two per week. In one week of 1872, almost a hundred vessels were lost or seriously damaged trying to make the fateful passage. A famous storm, October 16 of 1880, drove 30 vessels ashore in the Death Door area, including total loss of the Goodrich steamer Alpena, with a loss of 60 to 86 lives.[4] While some stranded ships could be salvaged, many were total losses due to the fatal damage done by reefs and rocky shores.

The Sturgeon Bay Ship Canal was built to provide a safe and sheltered passage between Lake Michigan and Green Bay. It was ceremoniously opened on July 4, 1878. Originally built as a private venture, modest tolls were collected from ships passing through it. In 1893 it was acquired by the federal government, and became toll-free. With safe passage through the canal available from July 1878, it is puzzling to read of 30 shipwrecks in Death Door Passage in the storm of October 16, 1880, and of many marine disasters here throughout the 1880s and the 1890s. One can only surmise that there were compelling reasons for ships to attempt the Door, or that thrifty masters preferred the risks of the Door to the payment of a small toll through the Canal. A later generation of ship owners continues to bellow their protests against St. Lawrence Seaway tolls, and so it may have been rather petty monetary considerations that took so many lives and so many fine ships on the rocky fangs of the islands and the shores of Porte des Morts.

Steam appeared on the Great Lakes very soon after the historic voyage of the Clermont on the Hudson River in 1807. But from the middle 1780s until late in the 1930s, sail first dominated the commerce of the region, and then survived concurrently with the side-wheeler and the propeller steamer, for about a total life span of 150 years. They sailed in numbers which cannot be precisely documented, but also in numbers almost beyond belief. Of simple but sturdy design, they were the product of craftsmen rather than of naval architects or engineers, and uniquely reflected the energies and the resources of the frontier country they served so well and so long.

C. The Age of Steam

Fulton's Clermont successfully paddled its way up the Hudson River in 1807. The age of steam-propelled vessels was not long delayed in arriving in the Great Lakes. On the Canadian side of Lake Ontario, the first paddle wheel steamer was launched near Kingston in 1816, and went into regular service early in 1817. She was the 740-ton 171-ft long Frontenac. In March 1817, on the American side of Lake Ontario, the 240-ton, 110-ft Ontario was launched. In 1818, a sister, Charlotte, was built for these waters. These ships, operating east of Niagara Falls in Lake Ontario, were of course confined to the lake and the nearby upper St. Lawrence River.

Steam then inevitably came to the upper Lakes. In 1818 the Lake Erie Steamboat Company was organized. Their first ship was named after an Indian chief whose name was translated as Walk-In-The-Water. In addition to being the first upper Lakes steamer, she was launched sideways, a method novel for that time but still followed today by many Great Lakes ship builders (Figure 2). Rigged as a two-masted schooner, she was 135-ft long and of 338 gross tons. The ship acquired still another unique distinction. Built at Black Rock on the Niagara

FIGURE 2. Side launch of a modern tuna clipper at Peterson Builders, Sturgeon Bay, Wisconsin. (Courtesy of Harmann Studios, Inc.)

River, the ship's engine could not produce enough power to move it upstream against the current, so on its maiden voyage it was towed up to the lake by 16 yoke of oxen, pulling on a hawser 300 fathoms long, which in turn was buoyed up on a bridge of a dozen small boats between the ship and the oxen on the river bank. The side-wheel steamer also rigged for sail was the conventional design on the lakes for several decades.

With the opening of the Erie Canal in 1825, a flood tide of immigration, both from Europe and the eastern U.S., poured through New York, up the Hudson, westward through the Erie Canal, and thence to the new lands in the west by paddle wheel steamers bound from Buffalo to Detroit and the new frontiers of Lake Michigan. There were no railroads, few families could undertake so long a journey overland, and so the common mode of travel was the steamer. On one major line, cabin passengers from Buffalo to Detroit paid $8 and deck passengers paid only $3.

The early steamers were not only package freighters, they were all-purpose vessels. One traveler's account describes a typical steamboat on Lake Erie as a "floating Babel", crowded with men, women, and children, speaking many languages; crowded with freight, baggage, and firewood for the ship's boilers; crowded with horses, mules, and oxen on the foredeck; and, for an early example of intermodalism, further crowded with wagons, coaches, carts, and conveyances of all descriptions. The early lake steamers hauled passengers and their belongings to the new western frontier lands and returned with the produce of the frontier. Manufactured goods from the East were in great demand on the frontier, and the settled

East was in need of the furs, timber, grain, and beef that flowed east and brought revenue and trade goods back to the westerly settlements.

In 1840 Captain James Van Cleve, a lake vessel master, while visiting New York City, was introduced to Captain John Ericsson, Swedish inventor of the screw propeller. Ericsson had not yet been able to persuade vessel builders as to the value of the propeller, and was greatly interested in the booming marine world of the lakes to demonstrate its merits. An agreement was reached, a company was formed, and in 1841 the Vandalia was launched. She was only 91-ft long and of 138 tons burden, rigged as a sloop. She was the first commercial propeller-driven vessel in the world, and thus the screw propeller proved itself on the Great Lakes. She was also the first lake vessel to have her machinery all the way aft, a design later to be standard for Great Lakes bulk vessels and many others. Still relying on sail as a standby, she was sloop-rigged.

For many years paddle wheel ships were called steamers while the screw-type vessels were called propellers. Both types coexisted with the fleets of schooners and other sail craft that poured out of lake shipyards. In 1843 the first "propeller" appeared on Lake Michigan. It was the Independence of 262 tons gross. It was also the first steam vessel to navigate Lake Superior.

Thus began the era of the "package freighter" under power on the Great Lakes, an era that was to endure for the century to follow, and largely to be terminated by changing technology and the effects of World War II. The sailing cargo vessel was a factor on the Great Lakes, as we have noted, from the 1780s to 1938, or for nearly 160 years. The package freighter's role was shorter, about a hundred years overall. As in the case of sail, there are regretably no comprehensive records of how many of the ships were built in total. From the 1880s onward, ships were formally documented, and there are historical records of the various fleets, as well as of individual ships, many of which had most colorful histories.

It should also be remembered that from the inception to the end, many of the lake package freighters were also passenger carriers. Pre-Civil War, and especially pre-railroad, they were the principal east-west carriers of the Great Lakes region. They carried cargo, cattle, and people with equal facility, and every conceivable item of commerce, moving between the manufacturing East and the pioneer West, made up their manifests.

The larger and more elite vessels, serving great numbers of passengers as well as varied cargo, had not only officers, engineers, and deck crew but also pursers, stewards, cooks, bakers, dishwashers, bellhops, porters, baggagemen, clerks, maids, laundry workers, and musicians. Some of the more palatial offered about the same amenities for a Great Lakes voyage, as did stately saltwater sisters of the transatlantic run.

The economics of this era of steamboating are of great fascination. They are exemplified by two advertisements which were carried in the Centennial Anniversary issue of the *Milwaukee Sentinel*, on July 4, 1876,[5] a Tuesday. The first, by the Goodrich Transportation Company, offered a round trip excursion, Milwaukee to Racine, by its steamer De Pere, leaving at 9:00 a.m. on July 4, and returning 8:00 p.m., for a round trip fare of $1. Again, for those who think the St. Lawrence Seaway route a modern phenomenon, another offer was truly striking. The advertisement was headed "Ocean Voyage To The Centennial" (at Philadelphia). The steamer Garden City would leave Chicago and Milwaukee "about the 15th inst." for Philadelphia "via River and Gulf St. Lawrence to the ocean," to remain 10 days and return. "Fare for the round trip, including 10 days on board at Philadelphia" was quoted at $150. The ship was described as "properly provided and fitted for the voyage". Since the round trip and the 10-day stop at Philadelphia must have considerably exceeded a month, this would appear to have been a rare travel bargain.

The 19th century was an era of teeming maritime activity throughout the Lakes region, with vast employment aship and ashore, engaged in building, repairing, operating, and servicing the fleets of sailing craft and steamers which carried the commerce. It was not

FIGURE 3. Whaleback freight steamer. First built in the 1890s this type vessel was the precursor of the modern Great Lakes bulk carrier. (Courtesy of the State Historical Society of Wisconsin.)

until the 1890s that the lake bulk freighter, the "whaleback", a small but important portent of the modern bulk carrier, came to these waters (Figure 3). From 1825 on, the century primarily belonged to the schooner, the side-wheeler, the propeller, and the passenger ship, and to the lake ports, large or small, which tied the region together with a flourishing water commerce of infinite variety and color.

D. The Railroad and Steamship Era

The 1850s brought the era of "trim and speedy railroad package freighters, seen and admired in every important harbor and on every channel of the five lakes".[6]

Between 1830 and 1850, a net of railway lines developed in the easterly part of the U.S., and by the 1850s, the first railroads began their extensions south, north, and west of the midcontinent rail hub soon to take form at Chicago. A number of eastern railroads extended lines to various easterly Great Lakes ports. Instead of surrendering freight to the railroads serving the western lands, the eastern railroads sought to keep control of the cargo, by a combination of rail movement in their own eastern territory, extended all the way to Lake Michigan and later to Lake Superior by the use of steamships owned by the various railroads. To accomplish this purpose, the eastern rails began to acquire both sail vessels and steamers. The system flourished, and within a decade the railway-freighter had become a major factor on the Great Lakes. The steamers they designed were "speedy for reliable performance, and capacious for efficiency".[6] By the turn of the century, 10 major east-west steamship lines were railway-owned, with 90 or more vessels.

The first rail-sponsored steamer service was the Union Steamboat Line organized in 1852. It started with 4 chartered vessels, then added 10 more charters. In 1854 it built 3 wood-hulled propellers; 3 more were built in 1856. Progressively larger vessels were built, or chartered, during the two decades 1860 to 1880. Among the numerous wooden vessels built was the Nyack, built at Buffalo in 1878. She was 231 ft long, 33 ft in beam, and carried 150 passengers. Her horsepower was a modest 750. In 1880 the line built the last two wooden hulls, 2,000-tonners. The line then went to a steady progression toward steel hulls and progressively larger and faster vessels. The first steel ship was the Jewett, built at Buffalo, 1882. This ship was 265 ft long and the gross tonnage was 1,953. By 1888 the line was building 340-ft ships. One of these, S. S. Chemung, set a record of 72 hr running time from Chicago to Buffalo. The last ships built for the line came out in 1896 and 1897. The fleet included two large and modern steel tugs, stationed at Chicago to tow Erie fleet vessels up and down the Chicago River.

The New York Central Railroad went into the lake business in 1855 with the incorporation of the Western Transit Company. Its fleet impressively numbered 6 schooners, 20 propeller steamers, and 200 canal boats. This fleet too was steadily enlarged and modernized. In 1873 it built the famous Arabia, the first iron ship, which was still in service over 70 years later. In 1893 steel construction began with the speedy Mohawk. Between 1898 and 1913, nine steel vessels were built for this fleet, five of them over 400-ft long.

The Pennsylvania Railroad's lake fleet was known as the Anchor Line, formally incorporated under the more unwieldy title "Erie and Western Transportation Company". It began in 1867 with a mixed fleet of a dozen, half steam and half sailing ships. Between 1876 and 1910, a total of 20 ships came off the ways for Anchor Line. Six were wooden hulls, two were iron, and twelve were steel. Noted for trim lines and an attractive color scheme, Dowling[6] describes the Anchor Line ships as the "most attractive on the inland seas". He goes on to say, "Even more appealing was their nomenclature, for after 1875 all of the ships of the line were named for the rivers of Pennsylvania, the beauty of which names has never been rivaled in Great Lakes history".[6] Among these were the famous Juniata, Tionesta, Octorara, Susquehanna, and other Indian names of great beauty. A wooden Juniata came out in 1878; a steel successor of the same name in 1905. In 1940 the Juniata became the then most modern and largest passenger vessel, as the Milwaukee Clipper, which operated cross-lake, Milwaukee to Muskegon, until 1970, carrying new autos, tourist autos, and some millions of people. During the summer tourist season a cross-lake volume of 250,000 people was typical for approximately a 4-month season.

Other railroad fleets took form. Northern Steamship Company, subsidiary of the Great Northern and Northern Pacific Railroads, was started in 1888. This line had two deluxe passenger ships and nine large package vessels, the last built in 1910. All ship names began with "North". The Lehigh Valley Transit Company, established in 1881, had thirteen steamers, six of wood and seven modern steel ships. Five, built between 1887 and 1890, were 4-masted, and were famous for fine appearance and speed, one running from the Soo to Detroit in a record 22 hr. Two ships built in 1901 were the largest package freighters, at 404.5-ft long and 50.5 ft of beam.

Lackawanna Railroad came to the lakes in the early 1870s, with five vessels. The Soo Line fleet, Minneapolis, St. Paul and Buffalo Transit Company, came on the scene in the late 1890s, also with five ships, all of Welland Canal size, about 3,000 gross tons.

The Rutland Railroad vessel operation, Rutland Transit Company, came from Ogdensburg to Chicago, via the Welland Canal route. Eight early wooden vessels came from the Detroit Dry Dock Company between 1883-1890. At that late date, they still carried auxiliary sail rig. Six steel ships were built 1907 to 1910. These ships carried marble from the quarries of Vermont to Chicago, and even such elite stuff as silks from France, transshipped from an Atlantic port. The Wabash Railway had the Lake Erie Transportation Company, with six ships, one of which had a life of well over 50 years, ending up as a Lake Michigan sand-sucker. Clover Leaf Steamship Line belonged to the Toledo, St. Louis and Kansas City Railway Company, operating nine charter ships.

Other package lines, some independent, some with railroad interchange features, were: The Union Transit Company, Port Huron and Duluth Line, and Chicago and Duluth Transportation Company. One Canadian package freight line operated from Georgian Bay port, to Milwaukee and Chicago, from 1890 to 1940, the exigencies of World War II taking its aging ships for other duty.

Table 1
FREIGHT TONNAGE THROUGH PORT OF MILWAUKEE PEAK YEAR (1913) AND SEVEN YEARS LATER (1920)[7]

	1913		1920	
Category	Net Tons	(%)	Net Tons	(%)
Across-lake	1,515,069	17.1	1,856,836[a]	31.7
Lower lakes	682,637	7.7	33,123	0.6
Along shore	701,266	7.9	246,010	4.2
All other (mainly bulk)	5,976,259	67.3	3,725,337	63.5
Total	8,875,231	100	5,861,506	100

[a] Increase principally car ferry traffic.

III. THE END OF THE PACKAGE FREIGHTER

A. The Panama Canal Act

In 1915 the Panama Canal, a dream since the days of Balboa, was completed and opened to world shipping. As expected, its effects on transportation were far-reaching. Railroad ownership of connecting steamship lines was well established in the coastwise and Great Lakes trade. The cross-continent railroad systems were of course powerful from the Civil War until the advent of truck, bus, and air competition. Congress feared that railroads also controlling intercoastal steamship lines might manipulate rates and shipping practices and thus nullify the effects of the new canal. In 1915 the Panama Canal Act became law, forbidding railroads to own steamship lines parallel to their roads, or to compete for traffic with water carriers they controlled. This legislation affected most of the rail-owned package freight lines on the Great Lakes, and forced the sale of the majority of the fleets as they then existed.

At Buffalo, in 1916, a group headed by a major stevedore contractor, William J. Conners, formed the Great Lakes Transit Corporation. It acquired a total of 36 ships from six railroad fleets, and established a major common carrier package service between Buffalo in the east, and Chicago-Milwaukee-Duluth in the west. It offered joint rail-water rates, with through bills-of-lading. It operated mostly at freight terminals owned by railroads as an adjunct to their former water operations.

In the post-World War I period, the Rutland-Lake Michigan Transit Company operated five ex-railway freighters between Chicago, Milwaukee, and the Atlantic Coast, via the Welland Canal-St. Lawrence River small locks. The last freight line of the package freight era was formed in the early 1920s, as the Minnesota-Atlantic Transit Company, or "Mat Line". Its first four ships were named Ace, King, Queen, and Jack, to which ships were later added with the names Ten and Nine. The Poker Fleet eventually numbered 12 ships. It disappeared in 1942 when its fleet was requisitioned for war service.

While rail control of water carriers may have been judged to create an undesirable interlocking or monopoly situation, the fact remains that in this era, lake ports teemed with traffic. Milwaukee Harbor recorded its all-time high for waterborne commerce in 1913, two years before the Panama Canal Act, with 8,875,231 net tons. Five years after the Panama Canal Act of 1915 the Port of Milwaukee commerce had decreased by 34% to 5,861,506 net tons (Table 1). Meanwhile rail traffic was increasing from 9,924,538 tons in 1911 to 13,407,299 tons in 1920.

Whatever the virtues of the Panama Canal Act, it is clear that for the large general cargo ports, and especially the east/west terminal ports of Chicago, Milwaukee, and Buffalo, the railroads, originally the sponsors and feeders of a vast lake trade, were forced into an adversary position vis-a-vis that trade. The trade continued on the basis of ostensible co-operation, with joint rail-water tariffs available, but it declined inexorably until its final demise in 1942.

Early in 1942, the U.S. War Shipping Administration issued requisitions for practically all vessels under the American flag useful to the war effort. Although few were designed for ocean service, they went to sea and were sent to many parts of the world on a variety of missions. Many were employed in the Pacific as interisland freighters. Some were lost in combat action; some were scrapped or converted in ocean ports to nonlake uses. The 50-plus ships involved have vanished as have untold numbers of ships plying these waters since the days of Cartier and Champlain.

B. The LaFollette Seaman's Act

Robert M. LaFollette, Sr., famed U.S. senator from Wisconsin, wrote, introduced, and fought to passage in the Congress, the Seaman's Act of 1912. It, like the Panama Canal Act, was to have profound effects on the Great Lakes steamship trade. Like the Panama Canal Act, it was intended to be protective, but its effects were destructive. It sought better and safer conditions for American seamen, primarily on the oceans. Especially on passenger ships, additional complements of offices, seamen, and steward personnel were necessary for compliance.

The Columbus was operated as a day-excursion ship, with no overnight passengers, but was nevertheless forced to increase crew and to reduce passenger capacity from 4,000 to 2,200. Her normal season was June 15 to Labor Day, with crew hired seasonally; the better men transferred to other year-round ships. The LaFollette Act required four full years of service to qualify a sailor as AB (able-bodied seaman). The rule forced the line to increase crew, thus increasing expense, and to reduce passenger capacity, thus reducing revenue. Elliott states, "The real decline in lake steamship operations began with the enactment of the LaFollette Act. Later some of the requirements were tempered, but not until much damage to lake shipping had been done."[8]

A special study by the U.S. Corps of Engineers provides a comprehensive review of Great Lakes transportation in the 1930s. Its coverage of the package freight trade as it then existed is revealing.

The long-term trend of Great Lakes package and passenger shipping was inexorably downward from 1920 on, due to four compelling factors: effect of the Panama Canal Act, railroad competition vs. earlier sponsorship, effects of motor car and truck competition, and obsolescence of ship types. Nevertheless, the decade of the 1930s was a decade of acute business depression, which led to many experiments and innovations in the effort to achieve the lowest possible shipping costs. Although many of the original basic commodities were beginning to decline, some substantial new traffic items had appeared, moving in large volume, such as newsprint, automobiles, steel products, canned goods, and sugar.

The New York State Barge Canal was actively employed at that period, and four motorship fleets operated low-clearance vessels of designs suitable for canal and lake operation. These carried considerable volumes of high-value cargo between Atlantic seaboard and Great Lakes ports, such as sugar, canned goods, auto parts, foodstuffs, and manufactured goods, generally moving in full cargo lots.

At midpoint in the decade, 1935, package cargo on the Great Lakes totaled well over 7 million tons of which 94% was domestic and the remaining 6% foreign. In that year, transport of trucks and autos by water carriers within the Lakes exceeded 461,000 net tons. Throughout the 1930s, autos were largely carried on decks of bulk freighters, and discharged over ramps built at receiving wharves. A few vessels were especially constructed to stow them between decks. With the passage of the Water Carrier Act by Congress in 1940, bulk carriers hauling mixed cargoes including autos would have become subject to Interstate Commerce Commission regulation as to rates and practices. They preferred to have fully exempt status as bulk contract carriers, and thus withdrew from auto haulage. The last commercial carriage of new autos ended with the retirement of the Milwaukee Clipper in 1970, and its sister ship, Highway 16 of the Wisconsin and Michigan Steamship Company, a veteran of the cross-lake service between Michigan and Wisconsin, in August 1973. Coincidentally, a Wisconsin port was the port of final sailing for the first ship, the Griffon; and Milwaukee was the port of departure for the final sailing of Clipper and Highway, which may prove to have been the last ships of the Lake Michigan package and passenger trade.

IV. HISTORY OF THE SEAWAY SYSTEM

The St. Lawrence route between the Great Lakes and overseas was not created anew with the opening of the present St. Lawrence Seaway in 1959. Small Canadian canals had circumvented the rapids of the St. Lawrence, and the predecessor of the Welland Canal had connected Lakes Erie and Ontario since shortly after the War of 1812, when Canada developed a route competitive with the Erie Canal which crossed New York State between the Great Lakes and tidewater. As early as the mid-1850s, wheat was shipped from the Upper Lakes to Europe. A series of consecutive enlargements of the route culminated, in the early years of the 20th century, in a number of canals along the present Seaway route. In 1932 the Welland Ship Canal, with eight locks, opened up Lake Ontario to the large "upper laker" bulk carriers. Subsequently, partially rebuilt, that canal constitutes part of the present Seaway. In 1933 the first regularly scheduled cargo liner service, utilizing small vessels, connected the Great Lakes with Europe. Such services, except for interruption during World War II, have been continuous, and after the opening of the present enlarged Seaway, larger vessels have been used for such liner services, in addition to numerous bulk-carrying tramp vessels. The depths to which the channels were dredged and the dimensions of the locks along the route, however, were much more influenced by the anticipated transits by lake-type bulk vessels than by the characteristics of salt water ships. The standard Seaway depth is 27 ft, and most of the major Great Lakes harbors have been dredged to the same depth, normally permitting access and transit by vessels of up to 25.75 ft draft. The results of these decisions, made several decades ago, is that a rapidly decreasing proportion of the world's oceangoing vessels can utilize the Great Lakes-St. Lawrence route, and the economies of scale currently realized on most of the principal ocean routes cannot be realized to the same extent for Great Lakes-overseas movements.

Until 1959, there were 22 small locks in the six canals bypassing the rapids of the St. Lawrence River in the 110 mi above Montreal. Lake vessels transiting to and from the lower St. Lawrence were designed to fit these locks. They were limited to a length of 259 ft, a beam of 43.5 ft, and a draft of 14.25 ft. With those dimensions, they could carry up to 3,000 tons of cargo. Over 200 such "canallers" were in operation through the St. Lawrence system during the period between World War II and the opening of the enlarged Seaway in 1959. Virtually all under Canadian registry, they included both "package" freighters and bulk carriers. The early pre-Seaway general cargo liners, which pioneered the Great Lakes-overseas direct trades between 1933 and 1959, were limited to the same dimensions, but because of their necessarily finer lines as seagoing ships, such "salties" could not move more than 1,600 tons on canal draft. They usually topped off with an additional 1,000 tons in the Lower St. Lawrence for the ocean voyage. Whether lakers or "salties," the vessels moved slowly through the canals and channels paralleling the St. Lawrence above Montreal. They commonly took 3 days between Lake Ontario and tidewater.

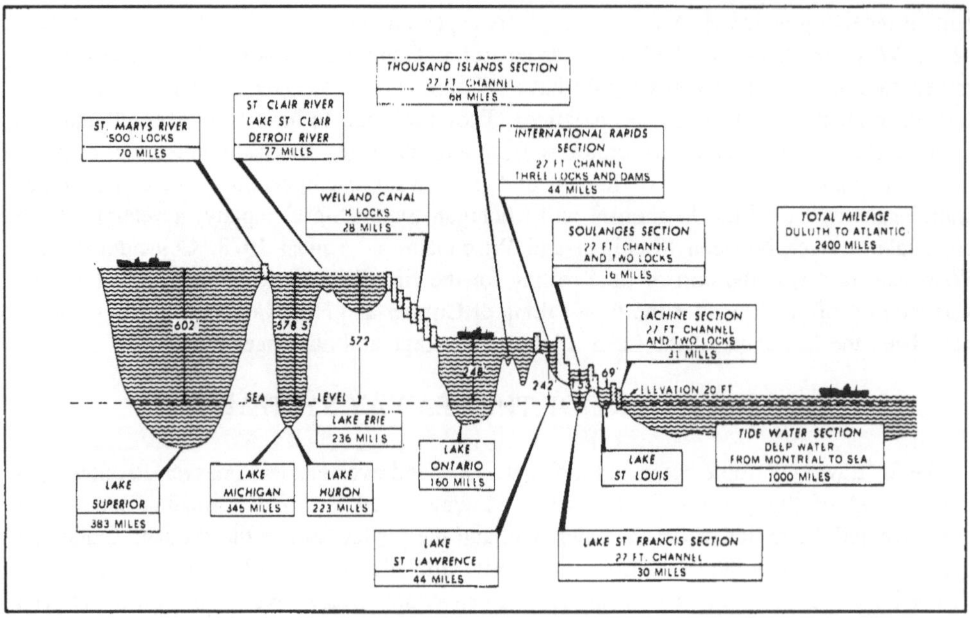

FIGURE 4. Cross section of the Great Lakes and St. Lawrence Seaway system, showing elevations and distances from Lake Superior to tidewater. (From International Great Lakes Levels Board, International Joint Commission, 1973.)

After 5 years of construction between 1954 and 1959, the Seaway was opened (Figure 4). The St. Lawrence portion involved the building of two dams: (1) a diversion dam at Iroquois, and (2) the Moses-Saunders Dam, with two million kw of electric-generating capacity between Massena, New York, and Cornwall, Ontario. Without the power, the Seaway as a navigation project would probably have been economically and politically unfeasible because the entire construction cost of facilities jointly used for navigation and power are charged against the power, rather than navigation. The costs and benefits originally were assigned for navigation on the basis of 29% to the U.S. and the remainder to Canada; later this ratio was changed slightly. Toll revenues are similarly assigned to the two nations. Of the seven locks between Lake Ontario and tidewater, two are within and were constructed by the U.S., in addition to the 10-mi Wiley-Dondero Canal.

Although most of the pre-opening traffic estimates for the Seaway were initially optimistic, during the early 1970s the traffic reached the volume projected for that period. However, during 1974 there was a substantial decline in traffic. As anticipated, throughout the entire 24 years of its operation the Seaway has been predominately an artery for bulk traffic rather than for general cargo. These bulk cargoes are carried in lake-type vessels, slightly modified for Seaway operation. Many of them were specifically built to the maximum dimensions of the Welland and Seaway locks: 730-ft long, 75 ft beam, and a maximum draft of 25.75 ft. With these dimensions they can normally carry up to about 28,000 tons of cargo. Because of higher costs of the U.S.-flag operation, the overwhelming proportion of such "maximum lakers" is under Canadian registry, although recent additions to the American-flag fleet are modifying that circumstance to an increasing extent.

A development of greater significance is the completion of the Poe Lock at the Sault Ste. Marie locks, which opened in 1970. In contrast to the other and parallel locks which limit the dimensions of the vessels operating between Lake Superior and the other Lakes to Seaway size, the Poe Lock admits vessels of up to 1,000 ft length and 105 ft beam; such lakers can carry over 57,500 tons at normal lake draft, or more than twice as much as any prior lakers.

Almost immediately, vessels of these dimensions were under construction. U.S. Steel's Roger Blough, built at Lorain, Ohio, was the first of the craft exceeding the earlier dimensions, although not of maximum possible length, and Bethlehem Steel's Steward J. Cort, assembled at Erie, Pennsylvania, became the first of the lakers at the new maximum size. Subsequently, other vessels of similar size were ordered, and some of the older vessels were enlarged. A new generation of lake ships is well under way. For the first time since 1932, however, a portion of the Great Lakes fleet is once again unable to operate east of Lake Erie because the vessels exceed the dimensions of the locks in the Welland Ship Canal as well as in the Seaway proper.

Except for the unusual circumstances of 1974, when strikes partially crippled the movements, a collision blocked the Welland Canal during the peak of the season and the business recession reduced industrial activity, it now appears that the St. Lawrence Seaway System will be limited, not by its ability to attract bulk cargoes, but rather by its physical capacity to handle the movements. In the early 1970s, the Lake Ontario-Montreal section handled nearly 60 million tons of cargo per year, very close to the early estimates of its capacity. Fortunately, the efficiency of the waterway route has been substantially improved by the trend toward larger vessels, handling more tonnage per transit; the "maximum laker" has superseded the very few surviving pre-Seaway small "canallers." Also, additional annual capacity has been provided by the successive lengthening of the navigation season, both with earlier openings and later closings, achieved by modern ice breakers and the installation of ice control devices in the locks themselves.

V. GREAT LAKES SHIPPING TODAY

The Great Lakes and St. Lawrence Seaway System is rapidly changing in significance as a major transportation resource. The changes are of substantial importance in terms of the need for reexamination and modification of planning and improvement programs; internationally, nationally, regionally, statewide, and locally.

The changes are economic, technological, and attitudinal. The roles of both internal Great Lakes transportation, and of Great Lakes-overseas transportation through the St. Lawrence Seaway have been subject to accelerating change in all those respects.

The internal Great Lakes transportation system is unique in its specialization and its adaptation to regional resources. Three dominant commodities, iron ore, coal, and limestone, are strategically located around the Great Lakes. They provide the genesis for the western hemisphere's largest concentration of iron and steel manufacture, and for other industries dependent upon iron and steel. The hinterland, both in Canada and the U.S., embraces some of the most productive grain lands of the world, important centers of food processing and grain milling, and areas of dairy production and meat packing.

The lake transportation system early tailored itself to these resources. Automation of bulk cargo handling is an old story to the Great Lakes. The self-unloading ship, considered a triumph of modern technology, appeared on the Great Lakes in the early 1920s, and the concept has moved forward consistently and efficiently for the past 60 years, climaxing with the new fleet of 1,000-ft vessels which are self-unloading (Figure 5). Similar automation was developed on shore, equally early, with the Hulett unloaders, giant coal bridges, and shoreside conveyor systems for bulk cargo. In a word, what's new on the oceans is two generations old on the Great Lakes.

Many of these systems came into being not only in response to unique resource patterns, intense industrialization of the region, and large-scale agriculture, but also as a necessary response to the seasonality of the system. The Great Lakes fleets were called upon to achieve delivery of a 12-month volume of materials in an 8 1/2-month shipping season. Thus emerged another characteristic of the system and the region: extensive shoreside stockpiling facilities, for iron ore and coal, grain elevators, and cement, stone, salt, and pig iron storage.

FIGURE 5. View of modern Great Lakes bulk carrier upbound with a cargo of iron ore. (Courtesy of Oglebay Norton Company, Cleveland, Ohio.)

From these circumstances there developed a transportation system which despite harsh climate and seasonal limitations, has quite often achieved system delivery in the range of 200 to 250 million tons/year (Figure 6). Within 10 years after the opening of the enlarged St. Lawrence Seaway in 1959, the Great Lakes-St. Lawrence System moved cargo in the range of 50 to 53 million tons/year, including about 15 million tons of cargo directly to and from overseas, with the remaining 38 million moving within the Lakes-St. Lawrence System, primarily in Lake vessels.

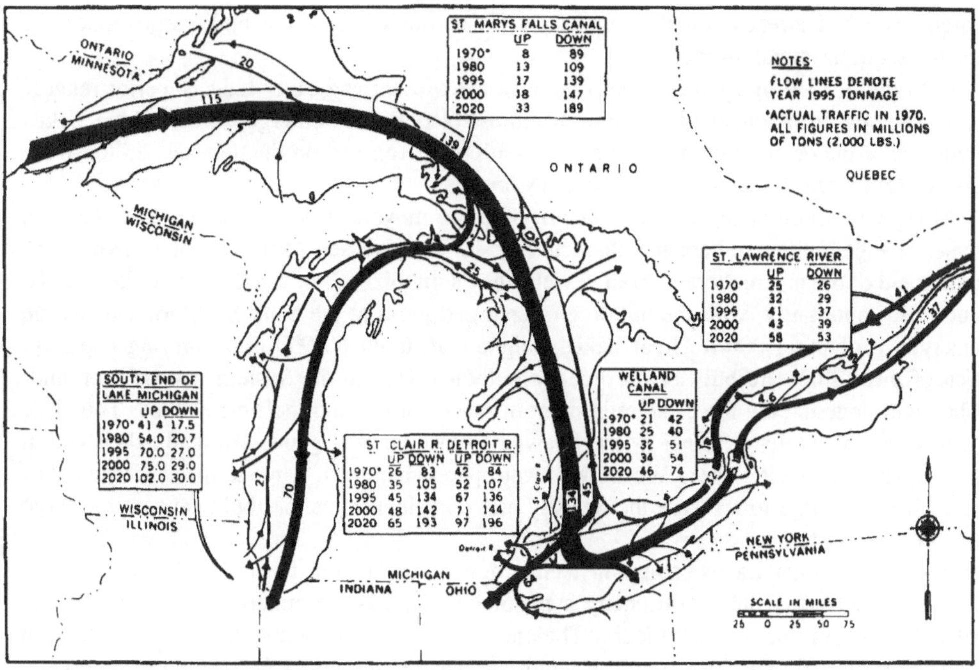

FIGURE 6. Great Lakes shipping tonnage, present and projected. Flow lines denote year 1995 tonnage. (From International Great Lakes Levels Board, International Joint Commission, 1973.)

The large lake bulk freighter, operated within the confines of the five Great Lakes, adapted itself to the new opportunities of the St. Lawrence Seaway, and with great flexibility engaged in the transport of grain from the western Great Lakes to St. Lawrence River elevators, and in the new westerly movement of iron ore from Quebec-Labrador back into the Great Lakes. In effect, the traditional pattern of iron ore from west to east was partially reversed, and an important new movement of iron ore developed from east to west, with grain, coal, and stone balancing out the vessel movements.

With little public awareness, there occurred one of the largest examples of block obsolescence of shipping in the entire history of the world. Prior to the opening of the Seaway in 1959, traffic between the Great Lakes and the St. Lawrence River had been limited to the dimensions of a 14-ft draft lock and canal system, capable of handling ships only 258-ft long and 42.5-ft wide. A fleet of about 250 of these "canalers" connected lake ports and the St. Lawrence River, busily transporting coal, grain, newsprint, and a variety of traffic between Lake ports and the St. Lawrence. With a maximum capacity of 3,000 tons, and crews averaging 35, the carrier fleet was obsolete overnight, when "upper lakers" of 30,000-ton capacity, and also with crews of 35, could deliver ten times the cargo volume per voyage. Within five years after the opening of the Seaway, the entire canal fleet of 250 vessels, practically all under the Canadian flag, had gone to the scrapyard, and the conventional Lake bulk freighter had assumed the role of these historic carriers.

The Great Lakes fleet between 1930 and 1975 shrank from 800 vessels to fewer than 300, but those 300 transport more cargo than the 800 did at their peak. In the Lake fleet, shipboard employment has shrunk from 25,000 officers and men, to fewer than 10,000. These trends are accelerating. The giant tanker on the oceans, another phenomenon of the last 20 years, demonstrated the economies of scale for other forms of bulk transportation. Dry bulk cargo carriers on the oceans moved rapidly from the 20,000-ton vessel to the 200,000-ton concept for coal, grain, fertilizer, iron ore, and other commodities. Great Lakes-Seaway general cargo trade had already felt the impact of this revolution, and the bulk cargo trades are facing a similar metamorphosis.

With the decline of open pit iron mining in the Mesabi and other Lake Superior ranges, and with cost inflation justifying new ventures, taconite has emerged in the Great Lakes region as a major commodity, while direct-shipping iron ore production has shifted from the western Great Lakes region to eastern Canada.

Along with these changes in the pattern of cargo movement within the Great Lakes is a major change in the numbers and characteristics of Lake vessels. These changes involve the retirement of old and moderate-sized vessels; the "jumboizing" of usable older Lake vessels; but most importantly, with the building of enlarged locks at the Sault St. Marie canals, the emergence of the 1,000-ft Great Lakes freighter of about 60,000 tons carrying capacity. Ten of these ships are built or on order, and their effect on the System will be profound. They will undoubtedly force the retirement of scores of old and inefficient ships. Too large to traverse the Welland Canal or the Seaway locks, the new giants are landlocked within the upper four Great Lakes. We now see the phenomenon of a modernized Great Lakes bulk fleet too large to traverse the Seaway and container ships and bulk carriers too large to enter the System.

Within the Great Lakes, shipping technology is responding to shifts of resources and industrial method. Coal, traditionally a major item in Lake commerce, declined with the advent of natural gas and petroleum. The energy crisis will undoubtedly require maximum utilization of coal. The historic fields south and east of the upper Great Lakes are already feeling the stimulus. Low-sulfur coals, found in the western states in huge deposits, are being commercially developed and are now moving easterly via Lake Superior ports to the Lower Lakes. Even as energy becomes a more critical question, the movement of petroleum continues to decline within the System, primarily shifting to pipeline. A Lake tanker and barge fleet of 97 vessels in 1960 shrank to only 39 vessels in 1974.

The advent of the container ship in the middle 1960s, and the second closing of the Suez Canal in 1967, accelerated a revolution in world shipping and transportation technology which has had reverberations within the Great Lakes as well. General cargo between the Great Lakes and overseas had reached a peak of about 8 million tons by 1970, and involved about 60 regular liner services. For about a decade, between 1959 and 1970, major lake ports also were fairly important ocean ports. They developed international trading complexes typical of successful ocean ports. Employment opportunities multiplied for the longshoreman and for all of the marine service trades. As transatlantic and transpacific container services matured, with large and fast ships, rapid port turnaround, and effective "land bridges" via the new interstate highway system and newly competitive rail intermodal system, the Seaway general cargo trade encountered severe competition. It declined in the early 1970s, as rapidly as it had developed in the early 1960s. In the middle 1970s only about a dozen direct overseas liner services were operating into Lake Michigan, and there was a parallel decline in port activities, terminal operations and employment.

With respect to both internal Great Lakes and St. Lawrence traffic, and direct Great Lakes-overseas ocean traffic, Great Lakes ports and shipping are losing much of their former versatility and diversification. The Great Lakes package freight trade disappeared after World War II except between Canadian ports, and that is rapidly declining now. A fleet of versatile ships handling automobiles, steel products, pig iron, machinery, and other high-value traffic, and capable of entering almost any port in the Lakes system, has largely been retired. The Great Lakes passenger trade, once flourishing, is virtually extinct. The Lake Michigan car ferry service, of major importance since the 1890s, is facing the competition of the "run through" train and modern highways. Two of the three car ferry routes have now been abandoned, and the third is facing the same fate.

The conclusion is inescapable that the Great Lakes-St. Lawrence Seaway is now concentrating largely on the mass movement of industrial raw materials (iron ore, taconite, coal and limestone) all fundamental to the steel industry along with grain, a major factor in the mid-continental economy in North American food supply to the world. The waterborne role of petroleum on the Lakes is rapidly declining, even as world use of this fuel has been increasing.

There is a growing tendency toward centralization of commerce at relatively few "load centers" on each of the seacoasts of the U.S. and other countries. The technology of the ship has run far ahead of the technology of the port. Ports are being required to respond to the need for deeper channels, to provide intermodal facilities which require large land areas, and to overcome partial or complete obsolescence of much of their physical plant. Inflation, high cost of equipment, and accelerating construction costs require massive amounts of capital. Many public ports are creations of local government entities which are struggling with environmental, social, and human demands upon their resources. Traditional governmental services suffer from the competition of needs or causes having higher social or political priorities. Thus local and private basis of financial support for port development may not be adequate.

It may also be questioned whether existing institutional structures responsible for Great Lakes public ports can respond adequately to many new developments: deconcentration of urban populations, inflation, shifting resources, new technology, the declining Seaway commerce, environmental challenges, and new "coastal zone management" concepts which will vitally affect ports on fresh and saltwater.

Despite these problems the Great Lakes-St. Lawrence Seaway transportation system remains dynamic and continues to evolve. The Great Lakes package freighters and passenger vessels of the past have been replaced by the giant 1,000 ft bulk carriers carrying coal and taconite, and the international grain traffic is taking on an increasingly important role. The Great Lakes transportation system and hinterlands remain a dominant industrial, urban, and agricultural force in North America, capable of playing a substantial role in the economic base of the continent.*

REFERENCES

1. **Barry, J. P.,** *Ships of the Great Lakes,* Howell-North Books, Berkeley, Calif., 1973.
2. **Wright, R.,** Center for Great Lakes Studies, Bowling Green State University, Ohio, personal communication, 1975.
3. **Heden, K. E.,** *Directory of Shipwrecks of the Great Lakes,* Bruce Humphries Publishers, Boston, Mass., 1966.
4. *Door County Advocate,* June 24, 1976.
5. *Milwaukee Sentinel,* Bicentennial Issue, July 4, 1976.
6. **Dowling, Rev. E. J.,** The vanishing fleets (The Great Lakes package freighters), in *Inland Seas,* Great Lakes Historical Society, January 1946.
7. U.S. Army Corps of Engineers, *Transportation on the Great Lakes,* Transportation Series No. 1, Revised 1937 U.S. Government Printing Office, 1937.
8. **Elliott, J. L.,** *Red Stacks Over the Horizon,* Erdman's Publishing Company, Grand Rapid, Mich., 1967.
9. **Schenker, E., Mayer, H. M., and Brockel, H. C.,** *The Great Lakes Transportation System,* University of Wisconsin Sea Grant College Program, Technical Report No. 230, January 1976, 292.

* Sections IV and V are drawn from Schenker, E., Mayer, H. M., and Brockel, H. C., *The Great Lakes Transportation System,* University of Wisconsin Sea Grant College Program, Technical Report No. 230, January 1976, 292.

Chapter 4

MAN UNDERSEA

Edward H. Lanphier

TABLE OF CONTENTS

I. INTRODUCTION

It is futile but fascinating to wonder who the very first human diver may have been. Was he (or she) a Neanderthal person or an early example of modern man? Did he (or she) make that first dive in emulation of a diving bird or mammal, or did it come about because some early swimmer saw and wanted something on the bottom and could not quite reach it from the surface?

A human diver trying to follow the example of a duck or a seal or a beaver must surely have been disappointed in his own performance. He could not stay down as long, or go as deep or as far, or find things as accurately; if the water was cold, he became chilled much sooner. Even though we have no certain knowledge about the earliest divers, we can infer much from studies of unaided divers in our own time. For example, the primitive ama divers of Japan and the hae-nyo divers of Korea have been studied rather completely in fairly recent years.[1] In them we can see clearly the basic capabilities and restrictions of the human body as a diving organism. With such a starting point, we can then see how diving apparatus and various innovations have been able to compensate for the handicaps of unaided human anatomy and physiology and how increased capabilities have in turn uncovered new limits and new problems.

II. HUMAN LIMITATIONS UNDERWATER

A. Breath-Holding Ability

The most obvious handicap of the human diver is the fact that he is an air-breathing mammal who largely lacks the special adaptations that enable other diving animals to surpass him as a diver. A human being can remain underwater without breathing apparatus only as long as he can hold his breath; and even with lifelong training, his breath-holding time remains only a few minutes at best. We breathe in order to take in oxygen and to expel excess CO_2. All reactions that liberate energy in the body ultimately require oxygen and produce CO_2; and in man, the use of O_2 and the production of CO_2 can be postponed to only a very limited extent. We can add to our CO_2 stores without serious harm, but the normal regulation of CO_2 levels is remarkably precise, and we ordinarily experience an overwhelming desire to breathe when the level rises much above normal.

There is only a small reserve supply of oxygen in the body, so we can interrupt the supply from the atmosphere for only a short time. Most of the tissues of the body can quickly shift to metabolic pathways that do not require oxygen immediately; but the brain cannot. There is no clear-cut warning of low oxygen levels in the blood reaching the brain, so loss of consciousness is often the first obvious consequence when the flow of oxygen from the lungs is slowed or stopped more than momentarily. Normally, it is the rise in CO_2 that tells a diver that he must come to the surface and resume breathing.

During exertion, as in almost any dive, the rate of O_2 use and CO_2 production is considerably increased, so the practical duration of breath-holding is even shorter than most of us can achieve during rest. Few unaided divers remain underwater for much more than two minutes. Hyperventilation (blowing off CO_2 from the body stores) can approximately double breath-holding time, but this is only at the risk of serious degrees of hypoxia (lack of oxygen) and possible loss of consciousness (see below).

B. Equalization of Pressure

The first time anyone tries to dive, he discovers that a surprising sensation of pressure develops in his ears. If descent continues, this is soon followed by pain unless he finds some way of *equalizing* pressure in his ears.

Normal barometric pressure at sea level is about 14.7 pounds per square inch (psi),

Depth (feet)	ATA	Partial Pressures (atmospheres) Assuming $N_2 = 80\%$ $O_2 = 20\%$		Gas Volume
0	1	0.8	0.2	1
33	2	1.6	0.4	1/2
132	5	4.0	1.0	1/5
297	10	8.0	2.0	1/10

FIGURE 1. Depth, pressures, and gas volumes. Several depths in sea water are used as examples to show important relationships in diving. The cylinders at the right illustrate compression and expansion of gas according to Boyle's Law. The density of the gas is proportional to the absolute pressure.

equivalent to the pressure exerted by a column of 760 millimeters of mercury (mmHg), 34 ft of freshwater, or 33 ft of seawater (Figure 1). We call this one atmosphere of pressure. It follows that the pressure surrounding a diver increases in such proportion as he descends. At 33 ft in the ocean, the total pressure will thus be twice what it was at the surface: two atmospheres, absolute (ATA). As he goes below the surface, the external pressure on the diver's eardrums increases while the pressure in the rigid air-containing middle ear space on the other side of the drum tends to remain what it was at the surface (Figure 2). The pressure-difference puts a strain on the eardrum. Meanwhile, pressure in all other parts of the body, including the lungs, airways, and blood vessels, increases along with the external pressure.

If the diver can coax his eustachian tubes to open, air from the upper part of his throat will enter the middle ear spaces and bring the pressure there up to the same level as the pressure in the water outside. (The same type of situation exists with the middle ear in flying or in rapid changes in height in the mountains or in tall buildings.)

An increasing pressure-difference across the eardrum can cause the drum to rupture. More often, the increasing external pressure makes itself evident in the blood vessels of the lining membrane of the middle ear. The result is swelling of the membrane and ultimately bleeding

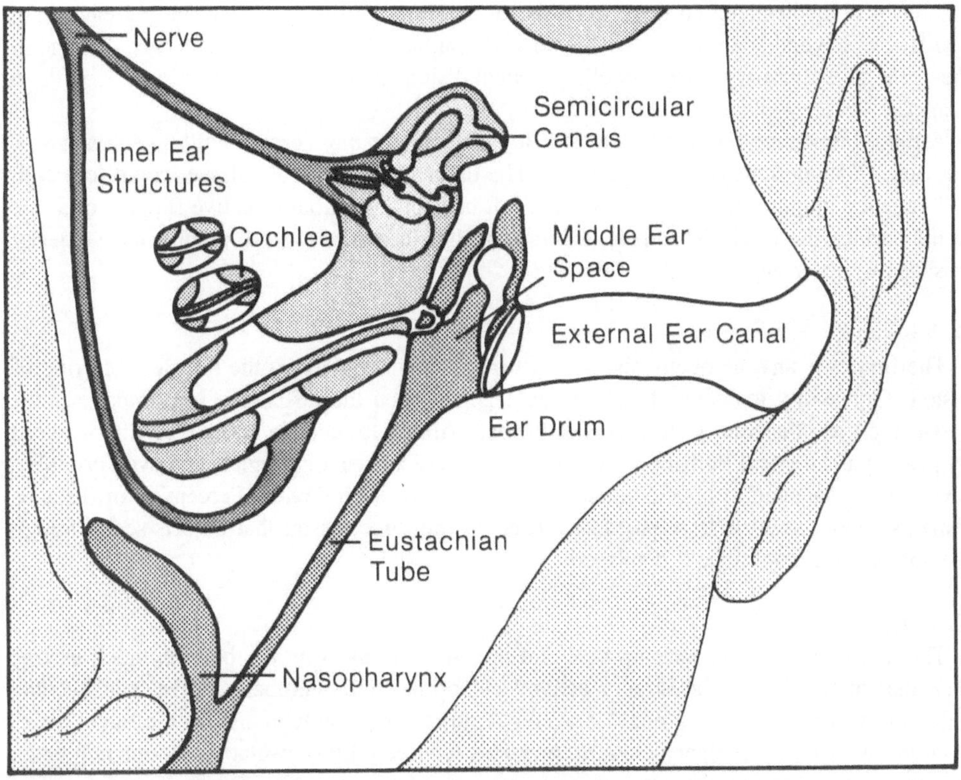

FIGURE 2. The human ear. The middle ear space contains the fragile bones of hearing. Pressure must be equalized in the middle ear during descent by passage of air from the nasopharynx through the eustachian tube. Otherwise, rupture of the eardrum (tympanic membrane) and/or other damage will occur. Adjacent structures are also shown.

into the space. Either may be sufficient to equalize pressure and spare the eardrum at the cost of other damage. The sinus spaces associated with the nose may fail to equalize if the membrane lining of the connecting passages becomes swollen. In an unequalized sinus, either massive swelling or "equalization by bleeding" is almost bound to occur and may lead to infection or other complications. The descriptive colloquial term for this mechanism is "squeeze."

C. Depth Limits

For most breath-hold divers, the depth of a dive is limited by trying to be sure to return to the surface before becoming desperately short of breath. Another kind of limit is imposed by the diver's anatomy if he attempts to descend to greater depth. This limit involves Boyle's Law, which indicates that the volume of a given mass of gas is inversely proportional to the absolute pressure (see Figure 1). If the diver started at the surface with a full inspiration, he might have a total volume of 6 ℓ of air in his lungs. He would ordinarily be able to exhale perhaps 5 ℓ of this volume; 1 ℓ would then represent the "irreducible" air content (residual volume) of his lungs. If the diver descended to 165 ft, the pressure around him would be 6 atm abs, and his original 6 ℓ of air would be compressed to 1 ℓ. Theoretically, he could go no further without physical injury to his lungs. In actuality, the "irreducible" lung volume is almost certainly smaller on compression than the residual volume as ordinarily measured after maximal exhalation on land. In addition, elevation of the diaphragm and movement of extra blood into the lung vessels and heart would make the "irreducible" volume still smaller. Actual depth records for breath-hold diving are currently greater than 300 ft. But beyond some point, shallower if the initial breath was less than full, damage in the form of pulmonary tissue swelling (edema) and hemorrhage would occur i.e., *lung squeeze.*

A more immediate problem related to Boyle's Law for most breath-hold divers is loss of buoyancy with compression of lung gas. The diver in the example above would lose 6 lb of buoyancy in going from the surface to 33 ft of depth. Without effective flippers or a line to the surface, he might be unable to return to the surface or even to combat the tendency to sink to greater depth.

D. Vision

The first time anyone opens his eyes in water, he is surprised at the hopeless blurring of vision. Our ability to focus objects in air depends upon the difference in refractive index between air and the watery structures of the eye. Ability to focus in water is readily restored by placing an air space in front of the eye by means of a mask or goggles, and even primitive divers have managed to provide themselves with some such device. Perception of distance remains abnormal: objects seem to be closer for the same reason that objects seen through the surface of a pool appear displaced.

E. Cold

The ability of water to take up heat and transfer it away from the body is much greater than that of air. Unless he is very active, an unprotected swimmer or diver can become chilled in water at a temperature that would be quite comfortable in air. By the same token, even the warmest of ordinary clothing provides relatively little insulation in water: it helps cut down heat loss only by preventing movement of water against the body surface. The primitive diver has a serious problem in cold water. His activity generates heat but at the same time promotes loss; and the same applies to the muscular activity of shivering. Animals and birds trap air in fur or feathers, and some animals can generate extra heat without muscular activity. The human body largely lacks this ability.

The fact that women have done most of the unaided diving in cold waters is surely no accident. The well-known effect of cold water on the male genitalia may help explain this division of labor; but females are genuinely better able to withstand cold because their insulating layer of subcutaneous fat is characteristically thicker.

F. Strange Maladies

Primitive divers probably came to terms with most of their limitations even though they did not understand them; but disasters that occurred only occasionally must have been terrifying as well as puzzling.

Two such calamities involve another gas law: Dalton's Law of partial pressures. This calls attention to the fact that the total pressure exerted by a mixture of gases is the sum of the pressures exerted by the individual gases present. From the physiological standpoint, it is the partial pressure of gases such as O_2, CO_2, and nitrogen that governs their effects.

In air, consisting of about 21% O_2 and 79% N_2, the partial pressure of O_2 is about 21% of the total (ambient) pressure (see Figure 1). When dry air is inspired, the partial pressure of O_2 is decreased not only by uptake of O_2 by the blood in the lungs but also by the presence of water vapor, CO_2, and usually a small excess of N_2. But when a breath-hold diver goes to depth, the partial pressure of O_2 (PO_2) increases along with the ambient pressure. If a diver could drop instantly from the surface to 33 ft, the PO_2 would essentially double. Because of this, a diver at depth can extract considerably more O_2 from his lung gas than he could at the surface. He may take advantage of this fact especially if he has depleted his CO_2 stores by hyperventilation: by huffing and puffing before the dive. The hitch comes when he finally realizes that he must return to the surface. As the ambient pressure drops, so does the PO_2. It may drop to the point where further uptake of O_2 by the blood is impossible and loss of consciousness from brain hypoxia becomes highly probable.

Even without the complicating factor of significant descent and ascent, many deaths have occurred from hypoxia, loss of consciousness, and drowning in breath-hold swimming or diving.[2] The usual victim is a healthy young male who is trying for greater distance in underwater swimming. He first hyperventilates vigorously, blowing off CO_2 and thus unintentionally ensuring that his PO_2 will be dangerously low by the time his PCO_2 builds up sufficiently to provide a warning. If observed, he may be seen to continue swimming under the surface in an increasingly erratic fashion; then he simply stops. Drowning follows unless he is rescued promptly.

Another catastrophe linked to partial pressure involves the increased N_2 pressure at depth. The increased pressure of gases in the lung causes the blood to take up unusual amounts of gas and to deliver it to the tissues. Oxygen is seldom a problem because the tissues are constantly utilizing it. A single breath-hold dive is not likely to deliver enough nitrogen to cause problems, but a sufficient succession of relatively deep dives with short times at the surface may cause a significant accumulation of nitrogen in the tissues. At some point, bubbles may form in the body on ascent, just as they do in a beverage charged with CO_2 under pressure when its container is opened and the pressure drops. It was only in the 1960s that the consequence of such bubble formation i.e., decompression sickness, the bends, was recognized as a possible complication of breath-hold diving.[3]

III. DIVING EQUIPMENT

Now that we have examined some of the limitations faced by primitive divers, we can see how diving equipment has been able to extend those limits. No equipment can repeal the gas laws or solve all of the problems; but extensions of capability have been very great.

A. Diving Bells

Early diving bells were basically just air-containing structures lowered to depth along with the diver. They were probably used in very early times, but reliable accounts date from the 1500s. A diver who can work with his head remaining in the bell has a reservoir of air that will become low in O_2 and high in CO_2 relatively slowly. If he must do breath-hold diving from the bell, it is as if the surface had been moved closer to the work site; but eventually the air will become foul as if he had been breathing it continuously.

The fascinating history of diving equipment[4-7] contains many devices that could scarcely have been practical. Some of these involve the principle of the "snorkel." Snorkels are very commonly used today because they permit the diver to keep his facemask in the water for underwater vision and yet breathe freely and comfortably. Usually, the snorkeling diver's back is barely awash. An attempt to use a long snorkel as a means of breathing at depth can be an educational experience. One soon learns that the pressure surrounding the chest can be an impressive force: inspiration is difficult even with the head of a standing diver barely submerged and it soon becomes impossible with further descent. (Dangers to the heart and lungs make this inadvisable as an experiment or demonstration.)

While some early inventions would have exposed the diver to "snorkel squeeze," others would probably have caused damage by attempting to house his head and chest in an enclosure connected to the surface while leaving arms and legs at the pressure of depth. Genuine progress had to wait for technological developments.

B. Air Pumps

Air pumps became practical around 1800, but reliable hoses may have been more difficult to come by. At any rate, the real milestone of diving equipment came with the inventions of Augustus Siebe in England. In 1819, he introduced what he called his "open dress." Here, the diving bell was shrunk down to the size of an inverted bucket with windows, ventilated by air pumped down from the surface. Attached to this helmet was a leather shirt. The excess air simply bubbled out around the bottom of the shirt as it would have escaped around the bottom of a ventilated bell.

In 1837, Siebe produced his "closed dress," which bears almost total similarity to the "hard hat" diving rig with which most diving was done until recent years. The conventional suit-and-helmet is still used for many operations. Powered air compressors are generally used, but manpowered air pumps can still be found. "Surface-supplied" diving gear now includes streamlined versions of the "hard hat" as well as equipment consisting of little more than a full-face mask and a hose. A nonreturn valve is essential to prevent the helmet or mask from suddenly being vented to surface pressure, a major cause of squeeze.

C. Self-Contained Equipment

Practical self-contained underwater breathing apparatus (scuba) also has a surprisingly long history. The oldest commercially available varieties were of the "closed circuit" type, where pure oxygen is breathed to and from a rebreathing bag with a canister of soda lime to remove CO_2. Such units are still used especially where silence and freedom from bubbles are a necessity. Only the volume of gas required by the diver's consumption of oxygen needs to be supplied, but the safe depth of use is very limited as will be seen.

The Cousteau-Gagnan Aqua-Lung® was the first really practical "open circuit" scuba. Its crucial component is a regulator (demand valve) that opens upon inspiration to give the diver the amount of gas he requires for each breath. He then exhales through an exhaust valve into the water. The gas is almost invariably compressed air, supplied from a high pressure cylinder or two carried on the diver's back. Many brands of open-circuit scuba are now available, but the basic principle and characteristics remain the same. Almost all sport and scientific diving is done with such equipment today.

Deep commercial work and some types of military diving require special gas mixtures, as will be explained. These in turn may justify the use of more complex equipment up to and including closed-circuit breathing apparatus containing oxygen analyzers and circuitry which adds oxygen as needed to maintain a predetermined partial pressure of that gas. Only with such refinements is it safe to rebreathe anything but pure oxygen in a fully closed circuit.

Open circuit equipment is extremely wasteful of gas: the volume required per minute is about 20 times the diver's oxygen consumption, and the amount required to ventilate the lungs during heavy exertion may be over 20 times that required at rest. Further, the need in terms of standard (surface) volumes is multiplied by the number of atmospheres of pressure at the depth of dive. In actual fact, the density of air at depth (see Figure 1) will generally limit the diver's ability to ventilate his lungs and will restrict his work capacity accordingly.[8] In any event, the useful duration of a practical self-contained gas supply becomes extremely limited with significant exertion in open-circuit equipment at depth. The more complex equipment permits relatively safe and economical use of special gas mixtures, but it is extremely costly to buy, and adequate maintenance is very demanding.

Even the simplest equipment has eliminated or ameliorated many of the "human limitations" that we discussed earlier. Breath-holding ability no longer sets any limits. Lung volumes no longer set limits to depth. Vision is rectified by glass view ports or face-masks. Cold exposure is tempered by air-containing "dry suits" or by the nonwetting aircells of foam rubber "wet suits." Various methods of positive heating have become practical or promising. The "strange maladies" have been replaced by less mysterious but no-less-daunting problems.

IV. OLD PROBLEMS AND NEW LIMITS

No form of diving equipment that leaves the diver exposed to the pressure of depth can circumvent certain basic medical problems of diving.[9,10] For example, the need to equalize pressure in the ears and sinuses remains, so a severe cold can render diving impossible despite the finest equipment. "Lung squeeze" is eliminated along with breath-holding by the availability of gas for breathing, but a new hazard is thereby introduced.

When a breath-hold diver ascends, the gas in his lungs cannot expand to a volume greater than the original lungful. Over-inflation or excessive pressure in the lungs as a result of ascent is thus exceedingly unlikely. Such is not the case when a diver has access to additional gas at depth. It might take only a brief period of breath-holding or a minor obstruction to cause a fatal accident on ascent if the lung is already nearly full. Regional gas-trapping could also be a problem. The most serious consequence is gas embolism, in which bubbles enter the bloodstream from the lung and then obstruct circulation elsewhere, especially in the brain. Such accidents are extremely rare when the diver is able to breathe normally throughout ascent. They are much more of a problem when he runs out of air or has a serious malfunction of his equipment at depth and must ascend without being able to breathe normally. As discussed below, prompt recompression is the only recognized treatment for gas embolism.

Being able to breathe at depth is a major blessing, but entirely normal breathing, especially during heavy exertion, is too much to expect. The density of the gas increases as it is compressed at depth in accordance with Boyle's Law (see Figure 1). At 99 ft, for example, air is 4 times as dense as it is at the surface, yet the diver must still take in about the same volume of gas in each breath, the same number of times per minute as he does when doing the same work at the surface. Otherwise, his CO_2 levels must rise. Some divers experience CO_2 retention because they habitually fail to increase their breathing sufficiently during exertion.[8] This tendency is accentuated by increased gas density even without the added resistance of tubes, valves, and canisters in breathing apparatus. Excess CO_2 can cause impairment of consciousness in itself as well as hastening the onset of oxygen poisoning and augmenting the effects of nitrogen (see below).

It may seem unlikely that oxygen is required for life but can also have poisonous effects, but this is important for divers to recognize. If oxygen is breathed for more than a day or so at a partial pressure equivalent to 0.5 atm or more, as with 50% O_2 at normal pressure, lung damage may ensue. If a working diver is exposed to a PO_2 close to 2 atm, as with 100% O_2 at 33 ft or 20% O_2 at 10 ATA, he may shortly develop a major convulsion like the seizures of grand mal epilepsy. Such an event at depth usually results in drowning.

The other main component of air is nitrogen, and this can also cause problems at depth. When air is breathed at 300 ft, where the pressure is about 10 ATA, the PN_2 is about 8 atm compared to the normal level of about 0.8 atm (see Figure 1). This can be evaluated in terms of an imprecise but instructive "gas law": Martini's Law. This states that the mental effect of each 50 ft of descent breathing air is approximately equal to that of one dry martini. The effects of N_2 are remarkably similar to those of alcohol. Not surprisingly, 300 ft is generally considered to be the absolute maximum of depth for diving with air.

As already discussed in connection with breath-hold diving, nitrogen is taken up by blood and tissues in an amount proportional to its partial pressure (Henry's Law). Increased amounts of oxygen are also taken up at depth, but the tissues are always utilizing O_2, while N_2 must come and go unchanged via the circulation and the lungs. If a diver has taken up a consequential amount of nitrogen at depth and then is returned to surface abruptly, bubble formation in the blood and/or tissues becomes likely. Some degrees of bubble formation are tolerated by the body, but beyond a certain point, or with bubbles forming in crucial locations, signs and symptoms will result. This form of bubble trouble is called "bends," decompres-

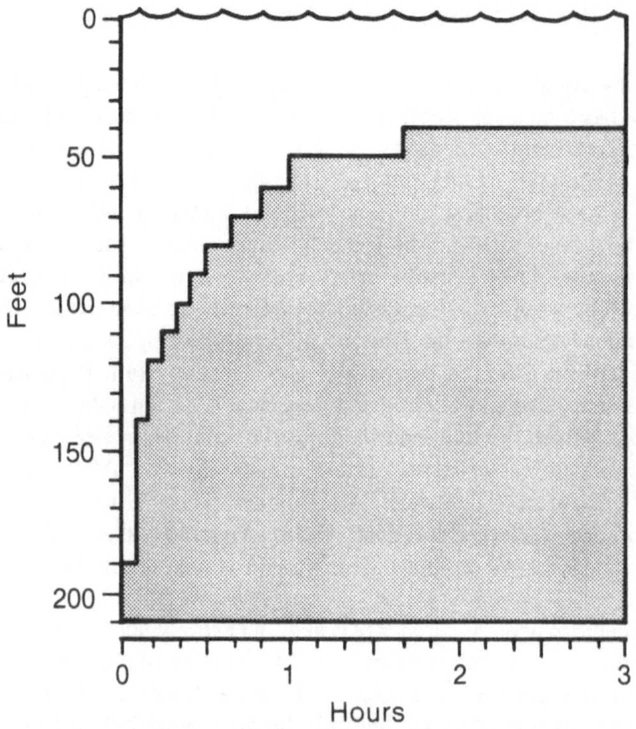

FIGURE 3. "No-decompression" limits. Decompression stops are not required on ascent according to the U.S. Navy Air Decompression Table if the "bottom time" at given depths is within the limits shown here (e.g., 60 min at 60 ft, 25 min at 100 ft). Bottom time is the elapsed time between leaving surface and beginning direct ascent at 60 ft/min.

sion sickness, or caisson disease. Most frequently, it is signaled by pain in or near a major joint. Less often but more seriously, the spinal cord or brain will be affected. Rarely, bubbles will obstruct circulation in the lung producing "chokes" and leading to circulatory collapse. The only effective treatment in most cases of decompression sickness is prompt recompression: placing the victim in a recompression chamber and increasing the pressure in the hope of compressing the bubbles to nonsymptomatic size and inducing the gas in them to return to the dissolved state. Recompression is also the only accepted treatment for gas embolism, but the bubbles in that condition (see above) have quite a different origin.

Prevention is preferable to treatment in decompression sickness as elsewhere. The condition can largely be avoided by limiting the uptake of N_2 by restricting depth and time. Within the so-called "no-decompression limits" (Figure 3), direct return to the surface is normally possible with little risk. Otherwise, a decompression table[11-14] must be followed. This prescribes stops at intervals on ascent to allow excess N_2 to leave the body without consequential bubble formation.

V. NEW SOLUTIONS AND NEW PROBLEMS

As has been seen, the limitations of the human body as a diving machine have been extended by diving equipment at some cost in complications, but the net gain in capability has been very great even with routine diving using compressed air. One of the fascinations of diving as a field of study is the ongoing effort to overcome successive generations of problems with new developments.

A dramatic example arose with the introduction of helium as a breathing gas for divers in the 1930's. Helium is only about 1/7th as dense as nitrogen, and it has no perceptible narcotic effect. It was thus able to break through the depth barrier imposed by nitrogen narcosis and the work limit imposed by increased gas density at depth. The greater capability was accompanied by some unfamiliar requirements. For example, the O_2 content of helium-oxygen mixtures had to be cut down to avoid convulsive toxicity and kept lower still to avoid lung damage in longer dives. Helium conducts heat away from the body much more readily than nitrogen, so the temperature of a helium atmosphere must be higher to avoid chilling. Helium also alters the sound of the diver's voice, so an electronic "unscrambler" must be used to render the diver's speech intelligible.

A new complication was discovered in the form of incapacitating neuromuscular effects when dives beyond about 600 ft were made with helium in the 1960's. "Helium tremors" were initially blamed on helium, as the name implies. Later, it became definite that it was not helium but the high pressure itself that was responsible. The name high pressure neurological syndrome (HPNS) was then applied. It includes not only tremors but also nausea and vomiting, "microsleep," and brainwave abnormalities. Convulsions have been induced in animals.

The severity of HPNS can be reduced by slowing the rate of descent and providing long stops as greater depths are reached. Even more dramatic is the effect of adding nitrogen to the He-O_2 mixture. For example, the record chamber dive to the equivalent of 2250 ft at Duke University in 1981 was made with "10% trimix", a mixture of helium and oxygen with 10% N_2 added. The N_2 not only offsets HPNS but also increases the density of the gas. At some point this approach will become impractical for that reason.

Already, several investigators have wondered what could be done at depths where helium itself would be too dense to allow the breathing required by useful exertion. The obvious answer lies in hydrogen, the lightest gas, half as dense as helium. With the low oxygen concentrations required at great depth, hydrogen is not explosive. It also differs from helium in having a mild but definite narcotic effect. Some authorities calculate that this might provide just about the right degree of narcosis to offset HPNS at any depth. If so, hydrogen would be the ideal gas for deep diving. At this writing, no one should dare to predict what the ultimate depth for man as a useful diver may be or how it will be reached.

One thing seems quite certain: no fancy gas or mixture is likely to get around the problem of decompression. The manifestations seem somewhat different with helium than with nitrogen, but the problem remains. It will surely be present with hydrogen.

One approach has helped to control the demands of decompression, and this is called saturation diving. Up to a point at a given depth, longer exposures require longer and longer decompression times. After two days or less, however, it appears that all of the body's relevant tissues have taken up as much extra gas as they can: they are now saturated at the depth concerned. Beyond this point in time, the diver can remain under pressure as long as he needs to without incurring additional decompression time. He can live in a habitat on the bottom (Figure 4) and work from it. He can commute to and from the worksite in a pressurized transfer capsule and live in a comfortable chamber on shipboard; or he may come and go in a specially-designed submarine while remaining under pressure. To avoid O_2 damage to the lungs, PO_2 must be kept below 0.5 atm, but this presents no great problem. The final decompression time will be long, but it will be much less than the sum of times for multiple short dives on most saturation jobs. However long, the decompression period can be spent in safe and comfortable surroundings.

A more futuristic hope is that of doing useful work at great depth without taking up excess inert gas at all. To date, this seems fully possible only if the diver remains at normal pressure in a small submarine with sophisticated mechanical arms or is encased in articulated "armor." Both of these possibilities are already practical for certain forms of work. But one can still

FIGURE 4. An undersea habitat. The habitat for the U.S. Navy Sealab III project is shown on a barge before being placed on the bottom at 600 ft. Note the figures in the foreground for scale. Ideally, the principle of saturation diving permits divers to live for weeks or months at the pressure of depth, then undergo a single period of decompression upon completion of their work. More commonly, divers live in a pressure chamber on shipboard and commute to the worksite in a pressurized personnel transfer capsule or bell.

wonder whether there will ever be a complete substitute for putting the diver's senses and dexterity immediately upon the work at hand; and that seems to require exposing him to the pressure of depth.

One intriguing line of thought appears to remain open: to accomplish the exchange of O_2 and CO_2 without involving a physiologically "inert" gas such as nitrogen, helium, or hydrogen. Such a gas is needed, basically, only to dilute oxygen and thus to keep its partial pressure within safe limits. In a gas, the total pressure must be accounted for by the sum of the partial pressures of the individual gases present. If O_2 represents a very small fraction, there must be a very large fraction of something else. In water at any ambient pressure, the partial pressure of dissolved gas can be as high as the total pressure, or it can be zero. A fish has no problem with O_2 toxicity at any depth because the water contains O_2 at only the same PO_2 as that of air at the surface.

A man cannot become like a fish, but there is reason to hope that breathing from a liquid-filled scuba might someday be made practical. There is much less chance that any "artificial gill" could ever permit direct extraction of sufficient O_2 from seawater. The most promising approach may be to pass some accessible fraction of the diver's total blood flow through a gas exchange unit outside the body: to add enough O_2 and to remove enough CO_2 so that the return of this blood to the main stream would yield the normal gas content of arterial blood. The technology for doing this already exists and is used in heart surgery. Doing it under increased pressure would reduce the fraction of blood flow required.

It really is too soon to predict what depth the human diver will someday reach, or how he will reach it!

REFERENCES

1. **Rahn, H. and Yokoyama, T., Eds.,** *Physiology of Breath-hold Diving and the Ama of Japan,* National Academy of Sciences, National Research Council (Publ. 1341), Washington, D.C., 1965.
2. **Craig, A. B., Jr.,** Summary of 58 cases of loss of consciousness during underwater swimming and diving, *Med. Sci. Sports,* 8, 171, 1976.
3. **Paulev, P.,** Decompression sickness following breath-hold dives, *J. Appl. Physiol.,* 20, 1028, 1965.
4. **Davis, R. H.,** A short history of diving, in *Deep Diving and Submarine Operations,* 7th ed., Saint Catherine Press, Ltd., London, 1962, 536.
5. **Dugan, J.,** The workshop of Hephaestus, in *Man Under the Sea,* 2nd ed., Collier Books, New York, 1965, 28.
6. **Marx, R. F.,** Diving machines, in *Into the Deep,* Van Nostrand, New York, 1978, 40.
7. **Vallintine, R.,** The earliest divers, in *Divers and Diving,* Brandford Press, Ltd., Poole, Dorset, England, 1981, 9.
8. **Lanphier, E. H. and Camporesi, E. M.,** Respiration and exercise, in *The Physiology and Medicine of Diving and Compressed Air Work,* 3rd ed., Bennett, P. B. and Elliott, D. H., Eds., Bailliere Tindall, London, 1982, chap 6.
9. **Edmonds, C., Lowry, C., and Pennefather, J.,** Physiology, in *Diving and Subaquatic Medicine,* 2nd ed., Diving Medical Centre, Mosman, N. S. W., Australia, 1981, 28.
10. **Lanphier, E. H.,** Medical aspects of diving and working in compressed air, in *The Merck Manual,* 14th ed., Berkow, R., Ed., Merck Sharp & Dohme, Rahway, N.J., 1982, section 255.
11. Navy Department, U.S. Navy Diving Manual, Naval Ship Systems Command, Washington, D.C., 1973.
12. **Miller, J. W., Ed.,** NOAA Diving Manual, 2nd ed., U.S. Department of Commerce, Washington, D.C., 1979, section 10.
13. **Shilling, C. W., Werts, M. F., and Schandelmeier, N. R.,** Decompression sickness, in *The Underwater Handbook: A Guide to Physiology and Performance for the Engineer,* Plenum Press, New York, 1976, 467.
14. **Empleton, B. E., Hill, R. W., and Lanphier, E. H., Eds.,** U.S. Navy air decompression tables, in *The New Science of Skin and Scuba Diving,* New Century Publishers, Piscataway, N.J., 1980, 240.

Chapter 5

WORK IN THE SEA

Robert A. Hamilton

TABLE OF CONTENTS

I. INTRODUCTION

Man works in the sea in variety of ways. The greater part of this is carried out on or from the surface. Some of these matters are covered in other chapters. This chapter deals with ways work is carried out with man working beneath the surface. Access to the sea is by means of various diving techniques whereby the worker is exposed to pressure, by submersibles and work stations kept at normal sea level pressure, and by remotely operated equipment and vehicles acting as surrogate workers. We take the position that the most interesting and important of this work is that done by diving.

It could be said that modern diving or the modern era of underwater work began with the loss of the nuclear submarine Thresher in 1963. It soon became painfully apparent to the U.S. that we had limited capability to search in deep waters and virtually none to rescue or salvage anything in great depths. This made the development of undersea capability something of a minor national effort which prompted many of the major corporations to invest in ocean oriented laboratories and created the concept that huge amounts of government funding would be available for this effort. The bubble didn't exactly burst, but before long it was clear that this was not going to happen; however, a great deal of interest had nevertheless been generated. The government development of the deep submergence rescue capability and Navy deep diving systems proceeded at substantial cost and questionable effectiveness, but commercial development of man's ability to work in deep water during that decade was phenomenal. The reason: petroleum.

America's hunger for energy had prompted the development of offshore oil fields (the first offshore well was drilled in 1947) and this called for deep diving. During the early 1960's commercial divers adapted mixed gas diving techniques using helium. Helium diving had been developed in the 1930's by the U.S. Navy, although by the 1960's the British Navy was also making significant progress and had shown a great deal more interest in deep manned diving than was evident in the U.S.

Perhaps the thing which made the early 1960's man-in-the-sea effort appear to be a new world was the development of the concept of saturation diving whereby a worker could remain for an essentially unlimited time at any depth which his equipment could support.

II. MODES OF UNDERSEA WORK

Human work underwater is clearly defined into two general categories, depending on whether or not the workers are exposed to the ambient pressure at the work site. Ambient pressure work is categorized as diving, and work in capsules and submersibles is called "one atmosphere" or "atmospheric pressure" or simply "atmospheric" work.

Diving is most generally categorized according to the type of equipment the diver wears, and this is roughly correlated with the depth and duration of the job and water temperature.[1,2] Beginning at the simplest level, considerable work is done underwater with breath-hold diving and minimal equipment, namely goggles, a means of buoyancy control, and possibly some thermal protection. The next level would be scuba diving whereby the diver breathes, by means of a regulator, compressed air or gas from a tank which he carries. Scuba is a former acronym which has matured into a real word: Self-Containing Underwater Breathing Apparatus. The scuba diver will have weights, a buoyancy vest, fins, and probably some type of suit for thermal protection, and will generally wear a face-mask instead of goggles. The mask covers the diver's nose so that pressure can be equalized inside the mask as he goes to greater depth. The scuba regulator which meters the compressed gas to the diver at a pressure approximately equal to that of his lungs is a "demand regulator"; it allows gas to flow when the diver lowers pressure in the mask or mouthpiece by inhaling.

The next level is surface-supplied diving.[3,4] In this mode the diver may use the classical old hard-hat gear or more modern "light gear." Hard hat gear has been described recently as "the kind of heavy helmet that you see made into lamps and things like that." Though there are many newer methods, this type of gear was introduced in 1859, it is effective for many types of diving and is still in rather widespread use.[5] The hard hat uses a flow-through breathing system whereby compressed air is pumped through the helmet at a rate fast enough to keep the oxygen level from being depleted and to remove the carbon dioxide exhaled by the diver. Some types recirculate part of the gas through a cannister that absorbs CO_2. This type of suit is comfortable in the water, gives the diver fairly good thermal protection, and the air in the helmet and suit gives him a few minutes of reserve breathing. However, it is heavy, cumbersome on the surface, and requires a good bit of assistance in order to get the diver in and out of the water.

The surface-supplied diver using "light gear" generally wears a full face mask which is equipped with a demand regulator quite similar to those used with scuba. The surface-supplied diver normally is tended from the surface, and in addition to the gas hose his umbilical contains a strength member, communications, and possibly a supply of hot water which is pumped through a special suit as protection from the cold. The surface-supplied diver might also wear a full helmet instead of just a mask. Such helmets are rather buoyant and thereby need to be heavily weighted, attached to the neck with some kind of yoke, or "jocked" to the diver's body with a harness. The surface-supplied diver also generally wears swim fins.

The next mode of diving is bell diving. A "bell" is a chamber holding two or three divers that is used for lowering the divers to the worksite and recovering them afterward; it also acts as a station for their work. The simplest type of bell stays at ambient pressure and may be a little more than a plastic bubble into which a diver can take refuge. The classical bell or submersible decompression chamber (SDC) is capable of withstanding external pressure to the depth at which the diver is working and may at times contain internal pressure (Figure 1). Virtually all bells can be mated to a deck decompression chamber such that divers can transfer between the bell and the chamber while remaining at pressure. In the classical bell dive the divers are lowered to the working depth with atmospheric pressure inside the bell. When they are ready to go to work the bell is pressurized to ambient pressure, and when pressure is equalized one of the divers gets out and performs his task. When work is completed

FIGURE 1. Diver exiting from diving bell. The diver's breathing, communication and hot water umbilical can be seen coming out of the bell hatch. This diver is wearing a typical demand-type "band" mask. (Photo courtesy International Underwater Contractors.)

(or time is up) the diver returns to the bell, seals it, and the bell is hoisted to the surface. In the meantime, the divers commence decompression by releasing the pressure in the bell according to a decompression table. The table prescribes the pressure and gas profile which the divers are to follow in order to be reduced to surface pressure without decompression sickness.

The diver who leaves the bell to work is called the ''lockout'' diver. The one inside is the tender, standby diver, or bellman. The lockout diver may use the same type of equipment described above for surface-supplied diving. The bellman is suited and ready to rescue the lockout diver.

III. UNDERWATER TOOLS AND TASKS

Whatever the equipment being used and more or less independent of the depth, the undersea worker has to have a bag of tools. This section discusses those items that are familiar and the same as used on the surface, plus a few that can only be used underwater.[7-9] Underwater techniques such as welding and photography are covered in other sections.

A. Diver Presence

A most frequent role of the diver in underwater work of all sorts can be summarized in just his being there. That is, the diver as an observer and supervisor of underwater work, whether it be automatic or controlled from the surface, is still a critical part of the operation. He is particularly useful as an observer when visibility is too low for effective use of television because his hands can help make decisions. The guiding of structures into place, opening and closing of valves, verification of the position of equipment and valves, reading of instruments, etc., are typical of the variety of tasks the diver performs. As in surface work, divers are often called on to do ''manual'' labor such as digging, cleaning, sandbagging, and so on.

B. Inspection and Maintenance

The degree of inspection and maintenance required for offshore structures came as something of a shock to oil operators in the North Sea when they had finished the construction of their production facilities and were ready to relax and enjoy the profits; it turned out that a great deal more ongoing work was necessary than had been estimated in earlier years. The importance of careful inspection and proper maintenance was underscored by the capsizing of the offshore platform Alexander Kjeland in 1979, which collapsed as a result of a crack in one of the legs.

A variety of techniques are used for inspecting undersea structures, and an intense research effort is continuously underway to find better methods and techniques. Inspection takes place at perhaps two levels. A welded structure, for example, when completed needs to be checked to make sure the welds are all sound, and then it will again need to be inspected periodically to check for cracks, corrosion, etc. Offshore divers have borrowed some inspection techniques from the nuclear industry; these include radiography, ultrasonics, and magnetic particle inspections.

The fundamental inspection method is visual. This has been done traditionally by a diver trained to know what to look for and how to report it properly, or by diver carrying a video camera with the essential judgement being in the hands of a topside engineer. More recently the camera is carried by remotely operated vehicle, ROV (these are colloquially known as RCV's (remote controlled vehicle) but ''RCV''® is a trademark). The ROV can handle a lot of inspections quite adequately, at much less cost, and much more quickly than with divers (see Section XIII), but there are also many situations where only a diver can do the job. One of these is when visibility is bad and the inspection becomes a tactile one rather than visual. This of course is much slower and less precise than visual inspection, but it works and often has to be done.

One interesting problem with inspection of the large offshore platforms is the diver's problem of keeping track of where he is. Operators have learned that it is necessary to put signposts all over the platform in order to guide the inspector or worker. One just cannot go around drilling holes and welding to a structure that is exposed to the special stresses of the deep ocean. Just figuring how to attach these signs was a significant problem that had to be solved. The platform builders also have had to develop three-dimensional coordinate systems and can now quite easily orient a diver or ROV with the better computer-based methods.

C. Hand Tools

Virtually all hand tools will function underwater, some better than others. As much of the hand work as possible is done at the surface before the structure it is put down; some work usually remains to be done underwater. Wrenches, crowbars, measuring tools, scrapers, and the like are routinely used there.

From these considerations it quickly becomes apparent that the best commercial diver is not necessarily the one with the best swimming ability or the one who can hold his breath the longest, but rather the one most comfortable in the environment and who can do the best work.

D. Power Tools

Underwater workers can benefit from the assistance of power tools in the same way as any other workers. These tools are used to perform such functions as jacking, chiseling, bolt tightening, cutting, operation of chainsaws and hacksaws, chipping, drilling, (both by rotary and impact methods) cutting and beveling of pipe (the beveling is to prepare the joint for welding), and so on.

There are three main types of power source used for these tools: pneumatic, hydraulic, and electric. Pneumatic power sources are limited to shallow water because they operate on a pressure differential and lose efficiency rapidly as depth increases. They are, however, relatively inexpensive and use a power source that is readily available at a dive site. Further, divers understand the handling of compressed gas. One type of pneumatic cleaning device uses captive needles, repeatedly driven against the surface to be cleaned.

Most underwater power tools are hydraulic. These may use either hydraulic fluid recirculated or open circuit seawater, but the latter imposes corrosion problems not yet completely solved. As with other engineering situations there are trade-offs that depend on the circumstances of the individual case; closed circuit hydraulic systems are more common. Hydraulic systems provide very high power and use relatively small and light tools, but they are relatively expensive.

Electric power tools are basically the least expensive and can deliver medium to quite high torque in small size using the new DC motors employing rare earth magnets. Underwater, however, electrical tools have one great drawback, the shock hazard. A variety of means can be used to make electric tools safe, including double insulation, ground fault detectors, and devices to interrupt the current when it goes through a path other than the intended one. Unfortunately, the alternating current that is so useful industrially is the worst risk from a shock hazard point of view. DC or very high frequencies (50 kHz) are relatively safe in the low voltage range, but AC in the 60 to 100 Hz range is too dangerous to be acceptable. For this reason, electric power tools are still rare underwater, although the means for making them safe are rapidly being improved, standards are being written, and their application is beginning to increase.

E. Jetting

Another effective technique is jetting, which is the underwater equivalent to hydraulic mining. Bascially it is little more than squirting water with a high-pressured nozzle to dig or loosen material or clean corrosion or marine growth off structures, etc. The jet loosens the material to be removed and may be taken away by an air lift (see next section). Pressures may be as high as 15,000 psi. Even at these pressures it may be difficult to remove hard calcareous deposits, so abrasives, cavitation bubbles, pressure intensifiers, and other new techniques are being investigated.

Jet nozzles are of a variety of sizes and shapes, some carefully engineered and others fabricated on the spot from available scraps of pipes. They share one common characteristic. If any significant degree of power is to be used, the diver cannot hold the jet against the reactive force. This is solved by making a T-shaped nozzle such that it points in both directions and the diver holds it at the point where the side arm tees into the nozzle bar. It is essential that the diver be situated so that there is no danger of his getting in the way of either of the jets because these can be extremely destructive at close range.

One specific problem with jetting is that it tends to stir up the bottom silt, and visibility may be reduced to zero when it is taking place. Also, divers may need to be protected from the sounds made by the high-speed water. Jets are also used on machines of various sorts, such as a device called a ''jet sled'' created to dig a ditch or cover a pipe.

F. Air Lifts

Divers use a clever means of moving mud, sand, silt, gravel or any other loose material from the sea floor to another location or to the surface. This is an air lift. The air lift is a pipe open at both ends, from 2 or 3 inches to over 1 ft in diameter and from several feet to many feet long, depending on what help the diver has in handling it, his supply of compressed air, and what he is moving. A short distance from one end of the tube a concentric nozzle pointed upwards (when the pipe is oriented vertically) releases compressed air. The air coming out of the nozzle creates a suction by Bernoulli's principle and the rising expanding bubbles entrain both water and the material being lifted. Since it acts as a vacuum cleaner the air lift is particularly useful in archeological work because it can remove sediment or sand without damaging whatever is being uncovered. Another advantage of the air lift over alternate approaches is that diving operations conventionally are equipped with a supply of compressed air and divers know how to handle it. Air lifts are often fabricated on the job by welding available scrap materials into the appropriate configuration.

G. Lift Bags

A lift bag is a plastic or fabric bag much like an enclosed parachute, which when filled with air provides buoyancy equivalent to the weight of the water displaced. Although it may resemble a plastic balloon, a lift bag has to be just as strong to contain air underwater as if it were hanging from a tree, full of water. Lift bags are designed to vent off the excess gas as it expands during ascent, in order to maintain constant volume and hence buoyancy.

Divers use lift bags to handle heavy materials and equipment as they would use a crane, and when many lift bags are ganged together they can be used to lift a sunken ship. Pontoons, injected foam, spherical floats pumped into the hull, and other such techniques are also used for raising sunken vessels.

H. Explosives

Explosives are eminently useful underwater, first in the traditional sense of blasting rock or structures but also in the more sophisticated use of shaped charges for cutting and in some cases even for welding. Special care needs to be used for several reasons. A shock wave travels much more easily through water than through air, and the potential of diver injury is great. Normally blasting is not done with divers in the water anywhere around the blast site. Underwater blasts can also damage other structures. There are techniques that can be used for shielding; one elegant one is the use of a curtain of bubbles as a means of attenuating a shock wave. Electric detonation can be used safely in remote areas where there is no construction, but in an area where there is a large amount of metal in the water, particularly where there are anti-corrosion currents or electrical equipment, it is considered unsafe to use electric blasting caps. Consequently, impact caps or caps set off with detonation cords are generally preferred.

Shaped charges have been used to perform some elegant cutting jobs underwater.[10]

IV. FISHING AND RELATED TASKS

From earliest time people have gone into the sea in order to harvest its bounty. Such work continues, and in many cases it is with the same techniques as have been in use for many years or even centuries.

Breath-hold diving accounts for a relatively small amount of work and production. The most widely known breath-hold divers are the Ama of Japan, (see Chapter 4 by Edward Lanphier) who harvest shellfish in water which may be as deep as 50 ft. In this operation the females do the work, largely because of their better tolerance to cold water.[11] There are tens of thousands of Ama at work in Japan and Korea every day the year around. They dive from a small boat at the surface, making up to 30 dives/hr to depths 10 to 25 m in water as cold 10°C. The diver uses a weight to speed descent, then is hauled to the surface by a "tender" who is usually the husband or another relative. The Ama usually wear goggles equipped with bulbs to prevent squeeze. Traditionally they have worn little clothing, but in recent years have begun to adopt neoprene foam wet suits.

Other mariculture, mostly in Japan, uses surface diving to tend and inspect nets, raise oysters, and to plant and harvest seaweed. Sport divers account for some harvesting, by spearfishing and related techniques.

A scuba rig gives a diver more depth and time capability, but relatively little production-type aquaculture uses scuba. It is, however, used extensively for the many inspection and maintenance tasks in work otherwise carried out from the surface. An example of this is the raising of salmon in pens in Puget Sound. The pens cannot be maintained in a functional condition without divers; though divers do not play a role in the day-to-day production, they are essential to the operation. Many undersea tasks are of this nature.

Surface-supplied diving is widely used for harvesting marine food and other biological products, as well as in aquaculture. The classical Greek and Turkish sponge diver and his hard hat diving suit still exists, but competition from synthetic materials and depletion of the beds ensures that this enterprise will not grow. This type of sponge diving was prominent in southeastern Florida until an epidemic wiped out most of the sponges. Scuba divers harvest coral in the South Pacific and Hawaii.

A popular and somewhat colorful enterprise, mainly off California, is abalone diving. Abalone are shellfish which may reach the size of a small dinner plate; the meat has to be pounded before being cooked, but it is a prized delicacy and very expensive. Though tightly regulated, stocks have become so depleted that relatively few divers make a living on abalone alone. It has been a good business in the past; quite a number of executives in today's major diving and diving-related companies got their start at abalone diving. Traditionally the diver worked out of a small boat equipped with a compressor and usually a tender, but some have been known to work completely alone. The successful abalone diver was the one able to find his product, which he did by walking or swimming along. He used a short steel pry bar (made from an old auto spring) and hauled his catch up in a wire basket. Sport divers with scuba account for most of the present abalone harvest.

Much diving in mariculture is research. This may consist of counting the catch in a gill net, or planting kelp "sporelings" on an open-sea frame, or work incorporating a pumping system for bringing deepwater nutrients up to the plants. Some planting and harvesting is done in the research area, but still most of the work is inspection and maintenance.

V. COMMERCIAL DIVING

The field of commercial diving could well be divided into two parts, work associated with petroleum, and everything else. Consider the "everything else" first.[12]

A. Construction

To a large extent diving construction work consists of most of the outdoor type work done in any construction, but done underwater.[4,6] It is of course less expensive to do as much of the fabrication as possible on the surface, just as it may be more competitive to "prefab" in ordinary construction. The diver then makes the final fitup, and he may end up having to do whatever needs to be done. Specifically this might be the setting and removal of concrete forms, making measurements, connecting structural members, laying pipe, and so on.

Much of the work is anything but exciting; it is grubby, cold, and uncomfortable, may require a lot of muscular exertion, and it is dangerous. The pay may be good, but it also may be quite modest, and the work is rarely steady.

As stated earlier, a good part of what a diver may do is just being there. He may guide the setting of a structure or the pouring of concrete, or he may be there to monitor something being done from topside. Divers and cranes work together frequently. Much work could be covered by the term "rigging," which means the handling of large, heavy items.

The construction diver more typically works in rivers, lakes, and harbors than in the open sea except in oil-related work.

Tools used underwater are essentially the same as those used at the surface, as mentioned earlier. Bolting or otherwise making connections or disconnections is a common function; it is done as often the hard way as it is with sophisticated power tools. This is partly economics, the cost-benefit tradeoff, but sometimes limitations on available capital leave little choice for the small contractor. Welding and some other specific operations used in construction are covered later.

B. Search and Salvage

One of the traditional concepts of what a diver does is that he finds things that are lost underwater, or at least looks for them.[13,14] Search and salvage may be recovering a lost fishing reel or outboard motor, but it may extend to salvaging lost submarines, sunken cargo, artifacts from the Andrea Doria, or Spanish treasure ships. There is glamour to some of this, but lifting ore from a sunken barge or raising a tug in New York harbor is more typical. One recent significant salvage job was the removal in 1981 of $90 million worth of gold from the SS Edinburgh sunk north of Murmansk in World War II. The special aspect of this exciting project was the fact that the water depth was 800 ft, and modern saturation diving techniques were used.

C. Ship Repair and Hull Cleaning

Maintenance of shipping accounts for a large fraction of the unglamorous end of diving. There is as yet no satisfactory method of keeping marine organisms, like barnacles, from growing on ship bottoms, and in the era of expensive fuel it is increasingly important to keep ships clean. This work is done primarily by divers using special equipment not unlike large underwater floor polishers. These devices operate with rotating brushes 2 or 3 ft in diameter and somehow are controlled, either by magnets, suction cups, thrusters, or buoyancy, such that the diver only need guide them. This operation is performed by military and commercial shipping enterprises alike. One technique used in this type of work on large ships is to begin the work on the bottom and work up the side so that any decompression obligation that might be incurred from the deep part of the job is taken care of during the work at shallower depths.

Another bit of maintenance that can be performed without a dry dock is the changing of propellors on all size vessels. This work, called a "wheel job," does not involve deep diving but on large ships does involve a good bit of sophisticated rigging and coordination between the diver and the topside workers. Ships have other equipment below the water line that often needs maintenance.

Additional ship support work by divers is involved in repairs of damage or other unscheduled maintenance below the water line. Temporary patches of several types are applied by divers and permanent repairs are made with welding and bolting methods, or the use of stud drivers. These devices use a cartridge equivalent to that of a hunting rifle and drive a captive bolt into a steel plate. The best ship repairs are accomplished by welding. High quality underwater "wet" welding is somewhat difficult, but still may be a cost effective approach (see Section VII).

D. Commercial Diving Training

A small industry has built up around the need to train professional divers. This is handled by both private vocational schools and public colleges. There have been for some years a few classical diver training schools which have provided the basics for shallow water "deep sea" diving. With the push into deeper water the need increased for divers trained in the special skills of operating with deep diving systems and with the special equipment required to do the job. This has caused several diving companies to set up schools primarily to train their own divers, but they also offer the training to others.

Military training has supplied a large number of commercial divers, but as commercial diving expanded and new techniques and requirements developed this source has become inadequate.

Regarding the training of divers, diving has always been an "apprentice" type industry. Until quite recently a young man who wanted to become a diver would work as a tender or deck hand on a diving operation for sometimes as long as 4 or 5 years before "breaking out" into becoming an actual diver. The tender in some respects is similar to the knight's squire, and in some situations is actually hired by the diver rather than by the company which hires the diver. The apprentice system worked and had its advantages, but was inadequate to meet the expansion and the high technology necessary for the 1970's; formal training schools were necessary.

It has been found by a study of the accidents and incidents in the North Sea by the British Department of Energy that human error is the single most important cause. This finding of course is not limited to diving. In order to improve the safety of North Sea diving operations the authorities have created requirements for competence as well as other more traditional things such as medical fitness in order for a person to be a commercial diver. In order for a diver to work on a petroleum related operation in the U.K. sector of the North Sea he must have a certificate indicating that either he has graduated from an approved diving school or has the equivalent in experience.

Meeting these training requirements is challenging, because of the equipment needed for the school and the water depth needed. In order to give a diver a dive to 300 ft it is necessary to take him far enough offshore to reach deep water. This requires a boat and a diving system, resources that are not available to most schools. Not only does equipment for this training require a substantial capital investment, but if the equipment were available it could just as well be used for regular diving jobs at a much higher profit level than when used for training.

One approach to meet this dilemma is the use of simulators. The diving environment can be simulated as effectively as those of the airplane or space craft, and one U.S. school has produced a simulation capability that can give the diver a large part of what he would experience if he were taking the actual training in deep water. The facility consists of a large pressurizable tank with a mock-up of a diving bell inside it. The divers are subjected to the pressure, light, thermal, gas, and of course wetness aspects of the actual working environment. The students control their pressurization in the mock-up bell. They are able to lock out into water at a pressure equivalent to that of 300 or more feet below the sea, in darkness if that is desired, and are able to work on typical subsea equipment. This operation is not without risk, but is far safer than actually doing it offshore and is much less expensive.

Another important aspect of the modern diving school is that in addition to teaching a "mode of transportation" (diving) it is necessary to train the diver-student in the use of the specialized equipment, particularly that required for inspection and nondestructive testing, "NDT." Photography, the operation of underwater television cameras, and use of other types of inspection equipment are all required. To emphasize this a page from the catalog of a diving school is reproduced as Figure 2.

VI. PETROLEUM RELATED DIVING

Offshore petroleum technology was fairly well developed by the late 1960's in shallow water but became highly sophisticated and well organized in deep water by the end of the 1970's; work at 500 ft is routine, and platforms have been set at 1,000 ft. During this entire period there has always been talk of "diverless" techniques for the many things that need to be done, and slow steady progress has been made in making this a technical reality. Even so, in the 1980's much remains to be done and it is not only necessary but economically advantageous to use divers for many of the jobs.[6,15]

Offshore petroleum operations can be divided into three more or less distinct stages. The first of these is exploration. This is the process of drilling the first oil well in a new area. This phase can be carried out without the absolute requirement for diver intervention but in the depth range where diving is economically practical, drilling is rarely done without diving support being instantly available. The cost of a shutdown is so great that an operator cannot afford to be without divers.

The next stage in producing oil is the production stage, which consists of tying the oil well in with the proper connection to a pipeline for collection of the gas and oil and transporting it by either pipeline or ship to market. Once the production equipment is installed the last work stage is that of continuing inspection and maintenance.

The vast amount of drilling up to the present time has been done with rigs or platforms based on the sea floor. Quite naturally the oil accessible in shallow water has been developed first. The stationary rig is generally a "jack-up" which might be a three-legged platform with adjustable legs so that the height of the platform above the sea floor can be adjusted to conform to the water depth. The equipment on this platform and on the other systems as well is essentially the same as that on a land based drilling rig.

Since most of the shallow water has been drilled, much of the activity nowadays involves drilling from a ship or semi-submersible drilling rig. These serve as floating platforms from which drilling takes place in much the same way as from the fixed platforms. They may be moored or maintained by dynamic positioning. The drill ships or "ship-shapes" are smaller and more vulnerable to weather and ocean conditions than the larger semi-submersibles and consequently require more diving services.

Once oil fields have been defined by the exploration process, platforms are installed from which additional wells are drilled and production implemented.

Course Outline—Professional Deep Sea Diver

	Approx. time (hrs.)	
1. Introduction	5	—
a. Introduction to course and regulations of PDS		
b. Course content and objectives		
c. Employment prospects for divers—present and future		
d. Individual career and progress advisement		
2. Underwater Physics & Physiology	10	—
a. Ideal gas laws—Boyle, Charles, Dalton, Henry Real gas laws—differential compressibility		
b. Pressure computation and air supply formulae		
c. Direct and indirect effects of pressure		
d. Breathing gas density problems		
e. Buoyancy—Archimede's Principle		
3. Medical Aspects of Diving	6	21
a. Red Cross First Aid & Cardio-pulmonary resuscitation training with final certification		
b. General Physical and Neurological exam.		
c. Prevention and treatment of diving related problems		
d. Environmental hazards		
e. Decompression related injuries and their management		
f. Intravenous and intramuscular injections in chamber		
4. Decompression Theory and Practice	14	16
a. Gas-tissue loading theory		
b. Decompression sickness—case histories and treatment		
c. Emergency recompression treatment tables, using chamber		
d. U.S. Navy air decompression and repetitive dive tables—Commercial application		
e. Exceptional exposure tables		
f. Surface decompression tables using oxygen and air		
g. Omitted decompression		
h. In-water decompression sickness management		
i. Mixed gas tables—U.S. Navy partial pressure helium-oxygen, helium, oxygen, nitrogen—N.O.A.A. air equivalent nitrox tables		
j. Medical lock operation		
5. Fittings and Hydraulic Principles	10	10
a. Design and application of commonly used regulators, valves and fittings		
b. Physical laws and design of gas and hydraulic circuits		
c. Chamber and bell piping systems design		
d. High pressure hose assemblies		
6. Communication Systems	2	5
a. Line pull signals		
b. Standard crane operators signals		
c. Visual and voice communications		
d. Basic electrical theory—trouble-shooting with multimeter		
e. Theory, design, use and maintenance of electronic communications		
7. Tending and Setting up dive station	2	5
a. Responsibility of tender in a diving operation		
b. Dive station set-up		
c. Procedures/records keeping		
d. Diver dress-in		
8. Working S.C.U.B.A.	3	10
a. O.S.H.A. and foreign regulations compliance		
b. Equipment—functional use and maintenance		
c. Working dives—inspection and construction		
d. Open water boat dives		
9. Surface Supplied Diving	—	47
a. O.S.H.A. and Foreign Regulations Compliance		
b. Theory of surface supplied diving		
c. Supply umbilical make-up		
d. Pneumofathometer systems		
e. High and low pressure breathing gas supply systems		
f. Gas consumption calculations and supply logistics		
g. Aquadyne mixed gas helmet and semi-closed rebreather system		
h. Spirotechnique EIP-5 push-pull system		
i. Swindell advanced helmet		
j. Aquadyne air helmet		
k. Desco helmet		
l. Kirby-Morgan band mask		
m. Aquadyne band mask		
n. Heliox-18 band mask		
o. U.S. Navy MKV deep sea outfit		
p. Constant volume suits		
q. Unisuits		
r. Hot water suit and heating system		
s. Working dives inspection and construction		
t. Open water boat dives		

FIGURE 2. Course outline for a diving school training commercial deep sea divers. This course lasts about 15 weeks. The left column shows "theory" or classroom time, the right "laboratory" or practice time, including diving time.

A. Exploration: Support of Rig Operations

If things go well the drilling rig may not need a diver at all. Typically the job a diver will perform on a drilling rig is dealing with something that has gone wrong. This ranges from adjusting the TV camera that is used to visualize the equipment on the sea floor, to recovery of dropped pipes or tools, or attachment of "guide wires." The oil well is drilled

KEY: Approximate time (hrs.)

☐ Theory

☐ Water / Practical

	Approx. time (hrs.) Theory	Approx. time (hrs.) Water/Practical
10. Power Mechanics Applicable to Diving	3	5
a. Gas/diesel and compressor trouble-shooting, maintenance and log-keeping		
b. Electric power generation systems		
c. Diesel/electric welding system		
11. Seamanship & Rigging	3	18
a. Nautical and navigation terminology		
b. Rules of the road		
c. Boat handling		
d. Rope knots and splices		
e. Heavy lift rigging		
12. Underwater Work Projects	—	110
a. Knots, bends and hitches		
b. Pneumatic saw, impact wrench, chipper operation		
c. Blind measurements		
d. Air lift, venturi and jetting operation		
e. Mooring cable installations		
f. High pressure pipe manifold assembly		
g. Gasket/"O" ring pipe connections		
h. Salvage 3 ton pipe section		
i. Marine structure inspection and repair techniques—Information recording and transferral to customer Pile coring—coupon sampling Magnetic particle flaw detection Ultra-sonic thickness—dye penetrant		
13. Cutting and Welding Principles and Their Practical Application	8	60
a. Topside and underwater theory		
b. Safety and emergency procedures		
c. Underwater welding of zincs, lifting-eyes and ground straps		
d. Production sheet pile cutting		
e. Oxygen/arc and thermite cutting devices		
14. Salvage Seamanship	4	7
a. Ship and oil rig construction		
b. Salvage law		
c. Damage control patches		
d. Methods of salvage		
e. Methods application to 3-ton salvage		
15. Mixed Gas Diving Principles	14	21
a. Purpose and theory of mixed gas operations		
b. Recirculators and demand principle systems		
c. Routine and emergency decompression procedures		
d. Management of "lost gas" and oxygen poisoning situations		
e. Mixing by partial pressure and mix-maker machine		
f. Gas analysis using galvanic, paramagnetic, N.D.I.R. and thermal conductivity analyzers		
g. Diving on He-O_2 and Nitrox Mixtures		
h. Gas control console systems		
i. Environmental control systems and CO_2 scrubbers		
16. Closed Circuit TV and Photography	7	8
a. Theory-operation and maintenance		
b. Underwater still photography		
c. CCTV internal pipe and hull survey inspection dives		
17. Bell Diving	3	21
a. Theory, principles, systems and procedures		
b. Control and operation of bell diving systems		
c. Working diver lock-outs		
d. Unconscious diver retrieval and resuscitation		
18. Offshore Oil Field Work	3	6
a. Introduction to oilfield terminology		
b. Diver's role in oil exploration and production		
c. Drilling systems operation and diver maintenance		
d. Replacement of guide wires, riser clamps, hoses		
e. Installation of flexible riser pipe		
f. Installation of rigid pipe hydrocouple		
19. Physical Fitness	—	30
a. Calisthenic drills throughout the course; open water swim training		
20. Safety	3	—
a. PDS safety regulations		
b. OSHA and Coast Guard regulations		
c. Association of Diving Contractors Safety Manual		
d. British, Norwegian and Australian regulatory agencies		
	100	400

*Sections 4, 6, 7, and 9 overlap each other so that actual practice time will be longer than those stated above.
**Additional Cutting and Welding will be part of Section 12.
***At sea and nighttime operations will be conducted throughout the course. In some cases, training will be in continuous shifts until the completion of a project.

FIGURE 2 (continued).

through a base plate, perhaps 20 feet square, that sits on the sea floor and is connected to the rig by cables at the corners known as guide wires. These are the first things to go if there is a distrubance, and they often need to be replaced.

The drilling rig working in deep water, up to 1,000 ft, normally has a deep diving system on board, and when working out from shore has divers on duty at all times. The divers go

FIGURE 3. Diver being lowered on stage near leg of semisubmersible drilling rig, giving an idea of the immense size of offshore equipment. This diver wears a ''bailout'' tank but receives his primary gas supply by umbilical. (Photo courtesy International Underwater Contractors.)

to work when needed and the diving company generally gets paid whether they work or not. The diver, however, is generally on a depth pay basis so gets more money if he makes a deep dive. For work in the Gulf of Mexico divers may be on a ''callout'' basis, and only go offshore when needed. The dive crew is entirely separate from the crew that operates the ships, which itself may be entirely separate from the crew that operates the ships, which itself may be entirely separate from the crew which drills the oil well. The economics of the offshore drilling rig play a big role in diving work. It is expensive to operate an offshore drilling rig; in today's dollars, it runs something around $100,000 a day. Thus if the dive crew can save even a few hours of time for the rig it can easily pay for its monthly service charges (Figure 3).

B. Completion

The next stage after exploration is the ''completion'' of the oil well. The work a diver does in completing an oil well is not in itself unique, but it deserves special mention because of where it fits into the scheme of offshore oil production. ''Completion'' is the step involved in connecting the oil well with equipment for separating oil, gas, and water, and for piping the oil either to shore or to a place where it can be loaded onboard a tanker. A ''Christmas tree'' is placed on the well, adjusted, and connected to a pipeline system that gathers oil from several wells.

The real significance of the completion is that this is the first stage in the production of offshore petroleum where the diver has been essential even when everything was going right. The work he does in the exploration phase is mainly to deal with problems, but most wells drilled today still require a diver to make the connection at the completion stage.

For the future there will be changes. The so-called ''diverless'' completion systems are making slow progress, and completions with the aid of a submersible or manipulator bell (see below) are becoming routine. Significantly, it has not been possible until the development of these systems for wells to be completed in water deeper than a diver can work (Figure 4).

C. Construction and Pipeline Work

The construction phase involves the most diver work. Drilling is done from either floating or jack up rigs, but production is from some type of platform. These have been placed in water as deep as 1,000 ft. Once the exploratory wells have defined the limits of an oil field the production wells may be drilled from a few platforms; often a dozen or more wells are drilled from a single platform.

The diver's work in platform construction is in many ways the same as steel work on the surface. The engineers attempt to plan everything so that as much of the work as possible can be done on the surface. The largest platforms have been towed to the site and lowered into place by management of their buoyancy.

The first deepwater platform was Hondo; it was assembled, then cut in two and reassembled as it floated horizontally near where it now stands in 825 ft of water off California. Shell's Cognac tower erected in 1978 to 79 in 1,025 ft of water in the Gulf of Mexico was built in 3 sections on shore and was then floated into place and assembled vertically, a section at a time. The work was accomplished without the absolute requirement for divers, although they were available and were used. At that time this was the deepest construction diving job yet performed.

Union Oil's Cerveza platform was built in one piece, loaded aboard a specially designed barge and towed from Morgan City, Los Angeles to its site in the Gulf of Mexico. Supported by pontoons along its eight legs it was floated into position by controlled flooding of the pontoons. It sits in water 935 ft deep; 34 wells can be drilled and produced from the platform over its 20-year projected life. Minimal diving work was needed.

Once produced and separated the crude oil is next taken to a refinery. The two main means of doing this are pipelines and tankers. If oil is loaded unto a tanker offshore, it will be done through some type of a loading buoy which moors the tanker and contains the piping and connecting equipment to load the oil into the tanker. The same type of buoy may be used at the other end to unload the tanker just offshore from the refinery or petrochemical complex for the next step in its processing. A large amount of diving work is involved in installing and maintaining these buoys.

Where it is feasible to build them, pipelines afford the best means of moving crude oil. These are connected to the separator complex and pumping facility and are laid on the sea floor following the most feasible route. As it is laid the steel pipe is coated first with a polymer or tar based "dope" and then encased in a reinforced concrete protective shell two or three inches thick. The pipe may or may not be buried, depending mostly on whether the area is used for bottom fishing and how much influence the fisherman have. Burying adds to the expense of building and sometimes of maintaining a pipeline but it considerably reduces the chances of damage.

A classical method for laying an underwater pipeline is with a "lay barge." These may vary from 100 to as much as 400 ft in length and are equipped to weld the pipe sections on to the existing pipe, section by section, and to lower the completed pipe to the sea floor as it is finished. Depending on pipe diameter and water depth this type of operation has been used for most pipelines.

An alternative to the lay barge is a reel barge, whereby the pipe is wound on a large reel and played out at the rear of the barge as it moves along. This type of barge is a great deal more expensive to begin with, but can lay pipe at a rather rapid rate. The protective coating is usually put on as the pipe is being laid. The largest reel-type lay barge, the Apache owned by Sante Fe Engineering and Construction Company, can handle 16 inch pipe.

FIGURE 4. Diving bell being lowered over the side. This bell mates on top of the deck chamber. This bell is guided by two guide wires leading to a large weight on the sea floor; its umbilical is handled separately from the lifting cable. (Photo courtesy International Underwater Contractors.)

Still another approach to pipe laying is to assemble the pipe in mile-long lengths on shore, tow it out, sink it, and connect the sections. The connections are made by mechanical joints or by welding.

In the lay barge operation the main work of the diver is to inspect the pipe as it goes out and as it settles on the sea floor, to insure that it is straight, i.e. without kinks, and possibly to assist in the process of trenching and covering it up. Where the bottom is soft the pipe is laid in a trench and covered, usually by a trenching machine or "jet sled;" such devices are operated and/or monitored by divers.

D. Inspection and Maintenance: Corrosion Protection

Protection against corrosion is a highly complex technology and involves divers at several levels. Their work is mainly in the installation of anodes and protective coatings, and in the cleaning and inspection of completed structures. Corrosion of steel in saltwater is an electrochemical process and it can be combatted rather effectively by means of applied electrochemistry, either by imposing appropriate currents counter to the corrosive action, or by means of sacrificial anodes. A sacrificial anode is made of a metal which has an electrochemical potential such that it will be attacked and corroded away before the steel structure. Zinc is the most practical for use on steel, and zinc anodes are placed on many structures. They are usually bolted to flanges which are welded to the structure; when necessary the anodes are replaced by rebolting or rewelding. In any case this job is invariably done by divers. Corrosion is often more of a problem near the surface, and a typical diver's job is to apply an epoxy coating to the "splash zone" of a structure. Corrosion is also much greater when the pipe is heated by the oil.

Pipelines also need continuous inspection and maintenance. These have protective coatings, and if damaged coatings are repaired in a timely way the pipe remains intact. When pipes are damaged they can be repaired by welding (see next section) or by external split-sleeve clamps. Or the damaged section can be removed and replaced using nonwelded couplings. One new idea for joining pipe uses a special nickel alloy that shrinks when warmed up from cryogenic temperatures.

VII. WELDING AND CUTTING

It is difficult to imagine the development of modern technology were it not possible to weld metals. Many welding and cutting techniques are used in underwater construction. This section discusses some of the more common applications.[5,6,8]

A. Burning

Burning is literally the correct term and is the term used by divers to describe the cutting of metal underwater. The fundamental process for cutting steel in normal surface construction is with an oxyacetylene torch. The flame heats the metal to the point where an additional flow of high pressure oxygen burns a slot by a combination of combustion and melting. Oxyacetylene methods are usable only to about 30 fsw (15 psi), because deeper than that acetylene mixtures become spontaneously explosive and are not safe to use. Hydrogen-oxygen has been used for cutting at depths deeper than that because it avoids this spontaneous explosion problem, but is not widely used at present. The primary method of cutting steel underwater is with oxyarc burning. The basic principle is to strike an arc the same as that of an arc welder between a hollow metal electrode and the work. High pressure oxygen blown down the hollow center of the electrode cuts the steel that is heated by the electric arc. The rod is consumed in the process and often needs to be replaced, but this is an extremely efficient and relatively inexpensive process.

There are variations on the oxyarc method. One of these is a "thermic" rod. If the material to be cut is not electrically conductive then an arc cannot be maintained between the electrode and the work. The method to make this possible utilizes a rod which itself burns and is consumed, but in so doing maintains the high temperature necessary to initiate the oxygen cutting. Like the oxyarc rod it is a hollow tube, but it is constructed of material which burns spontaneously once started, and it is then not necessary to maintain an arc between the rod and the work. This technique can be used for cutting wood, rusted metal, concrete, and similar materials.

B. Wet Welding

The correct materials and a certain amount of technique is required, but it is quite possible to maintain a welding arc underwater. The arc deposits metal from the rod onto the work. The particular problem of welding underwater is that the weld is immediately quenched by the water and this can result in crystalization of the steel and a brittle weld (Figure 5). This cannot be completely avoided, but by the use of special electrodes it is a feasible process. If a great deal of structural welding needs to be done it may be more economical to take the trouble to use another method, but for such things as the attachment of sacrificial anodes and construction repairs, and the like, it is worth performing wet welding. This method is in a continuous state of improvement.[16]

C. "Bubble" Welding

If the quenching of the hot steel by water is a problem the obvious solution is to weld in a gaseous environment. This is done by creating a gas bubble in the area of the weld. The diver is not inside the bubble, but rather it is transparent and he looks through it and does the welding inside with his hands. Because of the rather large forces required to maintain a gas bubble underwater it is expedient to keep the bubble as small as possible. These are generally plastic or otherwise light structures made to fit the specific application.

D. Dry Chamber Hyperbaric Welding

The most sophisticated and highest quality hyperbaric welding is carried out in seafloor chambers. These are called "habitats" by divers, but no one actually lives there. The highest quality underwater welding is done on pipelines. The consequences of a broken underwater high pressure oil pipeline are staggering, and the attention given by the owners and operators to the quality of the pipeline is in proportion to this hazard. Pipelines meet the qualifications of the American Petroleum Institute Standard 1104, whether at the surface or underwater. It is not yet possible to make such a weld in the wet, so pipeline welds are made in dry underwater chambers. Generally, the pipe is rigged into position by means of a pipe alignment frame or by careful manipulation of the pipe so as to align the ends. Whenever a joint is made, the chamber is put down over the pipe and sealed off where the pipe ends enter the chamber. The ends of the pipe are squarely cut and beveled and a "pup" piece is constructed to just bridge the gap between the two ends to be joined. It is not feasible to make the joint with only one weld because of the difficulty in aligning and cutting the pipes, but connecting larger pipes with a custom made pup is a rather routine matter. Normally the first or "root" pass is made with the TIG (tungsten-inert gas technique) which uses a tungsten electrode to heat the work and filler metal to form a bead much the same as traditional gas welding. With the inert gas shield and the right composition of electrodes an extremely clean joint can be made. Subsequent passes are made with appropriate "stick" coated electrodes.

Hyperbaric welders receive special training over and above their qualification as welders. The behavior of the arc and the weld metal are slightly different at pressure, and the process is sensitive to both the pressure and composition of the chamber gas. One welding category which has even higher standards than pipelines is welding for nuclear reactors, and welders qualified in that technique have made good hyperbaric welders.

FIGURE 5. Diver doing underwater "wet" welding. He is attaching a sacrificial anode to protect the platform against corrosion. (Photo courtesy of Sea-Con Services, Inc.)

Once a pipeline weld is complete it is necessary to X-ray it for certification. In its simplest form, this consists of attaching sheets of film to the outside of the weld and projecting gamma rays from a radioactive cobalt source located on the opposite side of the pipe. The shadow of the weld on the film discloses any obvious flaws and inclusions. More sophisticated welding assessment techniques resemble that of the "cat" (computerized axial tomography) scan used for visualization of brain structure. Welds must also pass hardness tests, and other inspection methods are often used at this stage. Once the joint passes its tests a protective coating is then applied.

Hyperbaric welding has been demonstrated at depths deeper than 1,000 ft of seawater but it is clear from that work that more research needs to be done to make this a routine practice.

E. Atmospheric Pressure Welding Chambers

One way of dealing with the difficulties of hyperbaric welding is to encapsulate the weld zone with a pressure-resistent chamber and evacuate it to atmospheric pressure. This requires some expense and engineering expertise, but permits standard techniques to be used for quality welds.

Whether hyperbaric or nonmobaric, one of the problems of seafloor dry welding is to provide the diver with a suitable breathing gas. Many jobs have managed this by having the diver breathe a pre-mixed gas with a full-face demand breathing mask. This method works and is safe, but it is somewhat cumbersome and expensive. Some companies that do a lot of hyperbaric welding have developed methods of cleaning the welding chamber atmosphere. Scrubbers remove the carbon monoxide, oxides of nitrogen, ozone and fumes which form as a result of the welding process, and the diver is provided with an essentially shirtsleeve atmosphere. He must of course wear eye protection.

The scrubbing of the welding chamber atmosphere is a good illustration of one of the economic facts of life that regulate underwater work. It is now and has always been a straightforward chemical engineering task to remove these contaminents, and to produce and monitor a safe breathing environment. However, partly because it represents an unknown area of research, and supported by the fact that the customer pays for the breathing gas, few companies have made the effort to develop the procedures. Companies that could have paid for this development in the breathing gas costs for one or two jobs have done dozens of them with the open circuit mask method and are even yet not inclined to change.

F. The Fire Risk

One might not think of fire as a significant risk to divers, but in fact quite of number have lost their lives to it. A chamber filled with compressed air, especially one with an enriched oxygen environment, supports combustion at a much greater rate than at the surface. If a fire does start the divers may be unable to escape from the chamber and may have no means of putting out a fire, so it is important to maintain fire safety precautions. These include the elimination of ignition sources, use of materials that will not burn in high oxygen environments, and the maintaining of the atmosphere as a noncombustible mixture. Also needed are adequate preparations for detecting and extinguishing a fire, and for providing emergency breathing gas by mask.

A relevant characteristic of the hyperbaric chamber atmosphere is that under pressures of 5 or 6 atm or more it is possible to have an atmosphere safe to breathe that will not support combustion. Whereas biological requirements for oxygen are satisfied in proportion to the partial pressure of oxygen, it is the percentage of oxygen that controls the combustibility of the mixture. Consequently, a mixture that is adequate to support life, a partial pressure of oxygen over 0.2 atm, will not support combustion if it involves a percentage of less than about six percent. This advantage is not available in normobaric welding, and fire protection needs to be provided in that circumstance.[17]

VIII. SCIENTIFIC DIVING

One might divide scientific diving into two catefories, that investigating the ocean and it's resources, and that investigating man's penetration of the ocean.[18,19]

A. Putting the Scientist in the Sea

The diving ocean scientist normally will be studying the sea floor, marine life, geology, pollution, and other environmental matters, or assessing resources. Most such investigators work from the surface, use scuba diving gear and can get along with relatively short bottom times. Those who need more time have the incentive for the use of sea floor habitats. These have been highly developed by the National Oceanic and Atmospheric Administration (NOAA) of the Department of Commerce and are described in the NOAA Diving Manual.[20]

B. Research on Diving Technology

Any aspect covered in this chapter may be the subject of research to improve it, and although this often takes place in simulators and pressure chambers, new developments ultimately have to be tested in the sea and a certain fraction of the activities there can be considered research and development.

Beyond research is the category of verification, validation or demonstration of a capability. While as of this time a few serious dives have been performed in the depth range 800 to 1000 ft, much activity in that depth and beyond has been performed primarily to demonstrate the capability. Pressure chamber work usually precedes work at sea; in 1981 chamber exposures reached 2250 fsw.

IX. MILITARY DIVING

There are two general categories of military diver, those concerned with special duties related to defense or warfare, and those who do much the same things as the commercial divers but in uniform. The U.S. Navy serves as the model, but the Coast Guard, Air Force, and Army also have divers.

The overwhelming proportion of the commercial-type Navy diving is in shallow water. A great deal is scuba, but surface-supplied equipment is widely used also. Harbor work, ship repair, inspection, and light salvage are typical jobs. Heavy salvage, at least in peace time, is usually contracted out to a civilian company. Nothing done by the Navy in this area differs from civilian work, but Navy equipment is somewhat different.[21]

Surface-supplied deep mixed gas capability is maintained, but rarely used for other than training. The Navy also maintains deep diving capability, usable for salvage or military missions in emergency, but these are primarily directed toward submarine rescue. The two Navy deep diving systems are on board submarine rescue vessels. The U.S. Navy has been a leader in development of both techniques and equipment, having been responsible for the saturation diving concept, for example, and several record-breaking deep dives both in pressure chambers and at sea.

While the Navy has led in development of new equipment and procedures, much of this development is not directly applicable to commercial jobs because of expense of the equipment and excessive manpower and logistic requirements of the Navy methods.

The other area of Navy diving is that directed toward military requirements. These include two main categories, "attack" divers and explosive ordnance disposal (EOD). The attack divers may belong to a SEAL (Sea-Air-Land) team, an underwater demolition team, or a similar group. Their missions may vary, but are generally clandestine and may involve launch and recovery from a submarine. These are the traditional "frogmen" that proved so effective in World War II. SEAL team divers may wear regular scuba gear in which case they make bubbles, or they may use a closed circuit system. The "rebreather" may use pure oxygen only, which limits diving depth to 25 ft, or may use an electronically controlled mechanism that provides the appropriate mix of oxygen and nitrogen (or helium, in rare cases). The possible missions of these divers are limitless. Because these breathing rigs allow long duration dives, special decompression computers are carried on some missions.

EOD divers work with unexploded ordnance, both ours and the enemy's, and their duty is to disarm or destroy it. EOD divers also clear mines, and for this mission wear nonmagnetic apparatus.

X. PHOTOGRAPHY

Underwater photography is both business and recreation.[22] Photography and its variations, e.g., video recording, are used for inspection in virtually all underwater work areas. This operation may be as a "live" inspection with a diver carrying an underwater TV camera under the direction of a topside engineer, or it may be a systematic video recording of the critical parts of a petroleum platform or a fish net. In all areas where visibility permits, photographs of the site are used in planning underwater jobs. The basics of photography are taught in diving schools along with welding and other work skills.

However, the overwhelming use of photography underwater is properly classed as "recreation." This involves the production of artistic photographs, books, films and slide shows as well as television. In fact, in many "salvage" jobs in the public eye, such as the many attempts on the Andrea Doria, the greatest profit lies in the television production, not the treasure, and many successful salvors take advantage of this phenomenon. Everyone is familiar with the exploits of Captain Jacques-Yves Cousteau, who is a respectable scientist (biologist, archaeologist, geologist) and also a superb showman and entertainer. It is hardly worth mentioning that it is television, not science, that pays most of the bills for the Calypso.

XI. THE SPORT DIVING INDUSTRY

Making it possible for divers and diving enthusiasts to pursue their sport is in itself a significant industry. It includes the production and sale of diving equipment, all aspects of travel, and a great deal of teaching. This latter area has expanded into the medical area in a curious way. Diving and hyperbaric medicine are covered poorly if at all in medical schools, yet a substantial number of diving accidents need to be treated, and a certain few disease entities can best be treated by hyperbaric oxygen therapy. The main ones are carbon monoxide and cyanide poisoning, gas gangrene, decompression sickness, and embolism. It is now possible for doctors to take postgraduate training in this specialty area in a warm-water resort area where diving conditions are good and training and supervision are available. Some dozen or so periodic courses are offered, and to a physician are tax deductible. These courses can be located through the Undersea Medical Society, 9650 Rockville Pike, Bethesda, Md., 20014.

XII. ONE-ATMOSPHERE WORK UNDERWATER

Diving is still the chief means of accomplishing underwater work, but as the depth increases the risk and cost of diving increases even more. This has prompted the development of other means of doing the work. Manned facilities not involving divers include submersibles and diving bells with manipulator arms, anthropomorphic diving suits, and stationary subsea work systems. These are referred to categorically as "atmospheric" systems. Remotely operated vehicles might form another category.

These underwater work systems are having an interesting effect on diving. As with any automation, there has been a fear that these would displace divers from their jobs. The overall effect has been probably in the other direction, for two reasons. First, the possibility of using diverless equipment makes subsea installation more attractive to the operators, and also in many cases these things have made diving safer. Perhaps not as much diving is done where atmospheric systems are available, but the need for divers is just as great as it always has been. Economy will ensure that divers will continue to work in shallow water because the high cost of the special equipment makes it economically more feasible in deep water where diving is particularly expensive and more hazardous.

A. Submersibles

As with most other sophisticated undersea work systems, the main user of manned submersibles is the offshore petroleum industry. Over 100 submersibles were at work, worldwide, in 1980.[23,24] The tasks performed include inspection, documentation, search an retrieval, debris clearance, and diver lockout support. Some have manipulator arms.

Launch and retrieval of submersibles in any but the nicest weather can be both troublesome and hazardous. The most widely used method is an A-frame on the stern of a work boat, possible in many cases to sea state up to 5 or 6. Other systems use rails, articulated and nonarticulated cranes, or cradles. A "moon pool" or centerwell is mainly used for diving bells.

For navigation submersibles use magnetic and gyro compasses, various types of sonar such as bottom mounted sonar transponder networks or sonar mounted on the support ship, and visual methods for tasks such as following a pipeline. Additional devices are usually carried for location after surfacing, and both floats and pingers are kept available for rescue purposes.

Life support duration varies from 17 to over 200 hr per occupant with the typical between 72 and 100 hr. Emergency food and water are carried on most submersibles, and extra thermal protection on some.

B. Undersea Subsystems with Manipulator Arms

As diving bells began to come into general use in the business of exploring for petroleum, it soon became apparent that a certain amount of work could be done by the divers in the bell without their actually having to lock out (Figure 6). Ordinary bells have been used for diverse types of inspection, monitoring of underwater work done remotely from the surface, and even for underwater searching. Often very simple tasks such as attaching a hook or looping a cable were all that was necessary, so even a primitive manipulator arm was effective in getting the job done without the need for exposing divers. Quite naturally there was an intense effort invested in developing underwater manipulators, and these efforts are paying off.

Modern manipulator arms have reached a high level of sophistication. These began with technology developed for the nuclear industry for the remote handling of radioactive materials, and they now provide motion on all axes, force feedback, that is, a sense of touch for the operator, and rather high levels of torque, grip strength, and weight lifting capacity. In many cases manipulator arms have specialized functions for close inspection, sawing or cutting, handling of special tools, etc. Manipulator arms have been installed on all varieties of submersibles, on standard diving bells, on bells having small thrusters that permit them limited motion about the work site, and on the more mobile diving bells, essentially tethered submersibles. One such bell is the Perry Oceanographic MDU (Mobile Diving Unit). The MDU has two chambers, with the one on top having room for three observers at 1 atm and one below for three divers. It has 2,000 lb of thrust and can work up to 600 ft from the support vessel. It has manipulator arms and a big clamp which can grab a platform leg and thus stabilize the MDU at a work site.

C. Anthropomorphic Diving Suits

The articulated, man-shaped diving bell deserves special mention. This system is maintained at atmospheric pressure, thereby enabling a man to work at even 1,000 ft without the need for decompression. The most popular such device is "Jim," which is shaped like a man and permits a diver to get around reasonably well at the work site. Other systems

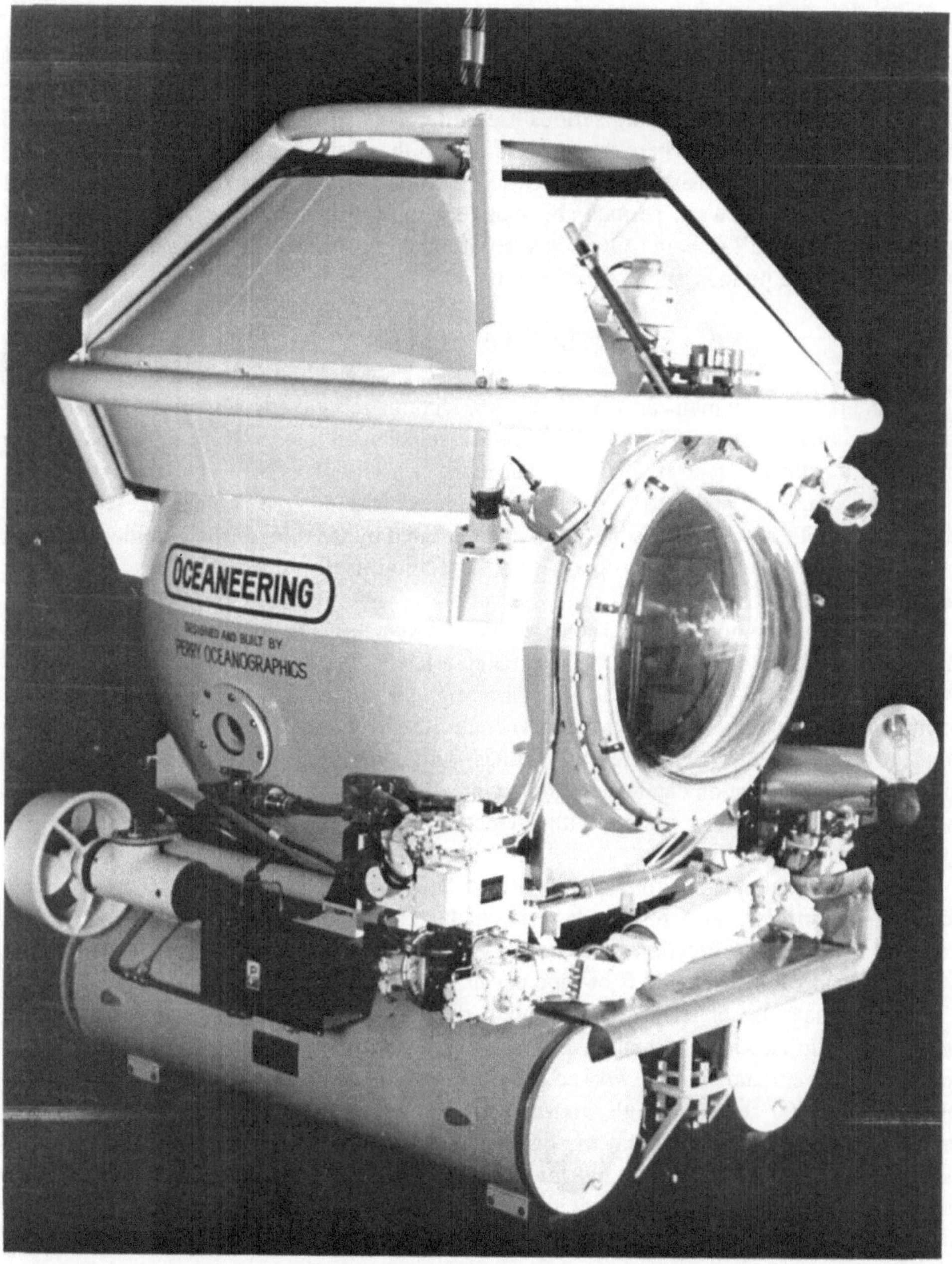

FIGURE 6. Manned observation/manipulator bell. This is a tethered but semi-independent bell which is not used for diver lockout. Some of these bells are capable of working as deep as 5,000 ft. The cylinders on the bottom are battery pods with 30 kWh capacity. This bell has thrusters (left) to provide worksite maneuverability, and can work from a "clump" weight on the sea floor to make it independent of ship motion. (Photo courtesy Perry Oceanographics.)

(Wasp, Mantis) are actually small one-man submersibles and rely on thrusters rather than walking for mobility.

The Jim operator is "hanging out" when he is at the end of an umbilical 1,000 or so feet long in total darkness, cold, and in general a most hazardous situation. Consequently, it is considered good practice to have another Jim suit or some other type of a vehicle available to support and provide rescue capability for the Jim that is doing the work.

D. Subsea Petroleum Hardware

As petroleum exploration moves into ever deeper water the need for completion and production capability without divers becomes increasingly important.[25,26] The first deepwater production systems were put into serious use in the late 1970's. Among these are a variety of subsea work chambers where the workers can do their jobs in a shirtsleeve environment. Most present systems consist of a small seafloor chamber to which workers can be transferred by submersible; their work periods are relatively short. Air is the gas breathed, and it is generally ''scrubbed'' of both CO_2 and hydrocarbon contaminants. Larger systems and those with longer work periods are in development.

XIII. REMOTELY OPERATED VEHICLES

One interesting way man can work in the sea without going there is by means of remotely operated vehicles. Over 400 remote controlled vehicles are presently in operation. These operate to depths as great as 3,000 m, but more typically at 1,000 m or less. All current commercial ones are tethered, although computer-controlled free submersibles are under development. To some extent small ROV's are limited by the drag of the umbilical. In high currents it is necessary for the support ship to position itself properly to allow the vehicle to fly to its desired location.

ROV's can be grouped into (1) tethered, towed vehicles; (2) bottom crawlers; (3) free-flying, tethered vehicles; and (4) the free-ranging untethered ones just mentioned.

The basic function of an ROV is to position lights and a television camera. The ROV is manipulated by thrusters (small propellors) controlled by a microprocessor, under guidance from a topside operator who flies the vehicle with a joystick that operates much like an airplane stick. Its location can be determined in part by the television picture, but for precise work an acoustic or sonar system with a video display shows the vehicle on a coordinate grid.

Some carry manipulator arms appropriate to their size, capable of only a few pounds of force on the smaller ones, but the larger vehicles can handle heavy objects. The bottom crawlers may perform functions such as burying pipelines. Brown and Root's ''MUT'' is 50-ft long and can bury 2,000 ft of pipe per hr.

One important use of ROV's is to support divers. The ROV follows the diver around, monitoring him and his equipment, and participating in the job at hand. One of the first jobs where divers and ROV's worked together was on the installation of Shell's Cognac platform at 1,000 ft in the Gulf of Mexico. Divers were apprehensive at first, fearing entanglements or other interferences, but soon realized the beauty and utility of the partnership. The ROV quickly acquired the name George, and soon the divers would not go out without him. An ROV also participated in the rescue of two divers in a lost North Sea diving bell in 1980.

XIV. ADDITIONAL READINGS AND REFERENCES

Relatively little of the development work on diving equipment and techniques is documented in the archival literature, but two sets of symposium proceedings hold a good sample of what has been done since the mid-1960s. The first of these are proceedings of a series of symposiums on equipment for the working diver held every other year from 1970 to 1978. These were sponsored by the U.S. Navy, held at Battelle Memorial Institute, Columbus, Ohio, and published by the Marine Technology Society. Another series of symposiums sponsored by the Association of Diving Contractors has resulted in symposium proceedings from 1976, 1979, 1980, 1981, 1982, and 1983. The addresses of these societies are: Marine Technology Society, 1730 M Street, N.W., #412, Washington, D.C., 20036 and Association of Diving Contractors, 1799 Stumpf Boulevard, Building 7, Suite 4, Gretna, LA, 70053.

REFERENCES

1. **Mount, T. and Ikehara, A. J.,** *Practical Diving: A Complete Manual for Compressed Air Divers*, University of Miami Press, Florida, 1975.
2. **Cayford, J. E.,** *Underwater Work: A Manual of Scuba, Commercial, Salvage, and Construction Operations*, 2nd ed., Cornell Maritime Press, Cambridge, Md., 1966.
3. **Hampton, T. A.,** *The Master Diver and Underwater Sportsman*, Arco Publishing, New York, 1970.
4. **Martin, R. C.,** *The Deep-Sea Diver: Yesterday, Today and Tomorrow*, Cornell Maritime Press, Cambridge, Md., 1978.
5. **Thompson, F. E.,** *Diving, Cutting and Welding in Underwater Salvage Operations*, Cornell Maritime Press, Cambridge, Md., 1944.
6. **Zinkowski, N. B.,** *Commercial Oil-Field Diving*, Cornell Maritime Press, Inc., Cambridge, Md., 1971.
7. **Hackman, D. J. and Caudy, D. W.,** *Underwater Tools*, Battelle Press, Columbus, Ohio, 1981.
8. Marine Technology Society, Proc: Symp. Underwater Welding, Cutting and Hand Tools, Marine Technology Society, Washington, D.C., 1969.
9. **Haux, G.,***Subsea Manned Engineering*, Best Publishing, San Pedro, Calif., 1982.
10. **Rynecki, A.,** Salvage of the tankers "Elias" and "Corinthos," in, *The Working Diver 1976*, Marine Technology Society, Washington, D.C., 1976.
11. **Rahn, H. and Yokayama, T.,** Physiology of breathhold diving and the Ama of Japan, *Natl. Acad. Sci./Natl. Res. Council*, 1965.
12. **Submex,** *Professional Diver's Handbook*, Submex, London, 1982.
13. **Brady, E. M.,** *Marine Salvage Operations*, Cornell Maritime Press, Cambridge, Md., 1960.
14. **Gores, J. N.,** *Marine Salvage*, Doubleday, Garden City, N.Y., 1971.
15. **Kash, D. E., White, I. L., Bergey, K. H., Chartock, M. A., Devine, M. D., Leonard, R. L., Salomon, S. N., and Young, H. W.,** *Energy Under the Oceans: A Technology Assessment of Outer Continental Shelf Oil and Gas Operations*, University of Oklahoma Press, Norman, OK., 1973.
16. **Helburn, S. and Reynolds, T.,** Recent advances in wet welding to 300 fsw (90 msw), in, *International Diving Symposium '82*, Association of Diving Contractors, Gretna, La., 1982.
17. **Shilling, C. W., Werts, M. F., and Schandelmeier, N. R., Eds.,** *The Underwater Handbook: A Guide to Physiology and Performance for the Engineer*, Plenum Press, New York, 1976.
18. **Woods, J. D. and Lythgoe, J. N.,** *Underwater Science*, Oxford University Press, London, 1971.
19. **Somers, L. H.,** Research diver's manual, Technical Report 16, University of Michigan-Sea Grant Program, Ann Arbor, Mich., 1972.
20. NOAA Diving Manual, 2nd ed., Natl. Oceanic Atmospheric Administration, Washington, D.C., 1979.
21. U.S. Navy Diving Manual, NAVSEA 0994-LP-001-9010, including changes 1 and 2, Navy Department, Washington, D.C., 1979.
22. **Frey, H. and Tzimoulis, R.,** *Camera Below*, Association Press, New York, 1968.
23. **Busby, R. F.,** Manned Submersibles, Office of the Oceanographer of the Navy, Washington, D.C., 1976.
24. **Busby, R. F.,** Manual of Manned Submersibles: Design, Operations, Safety and Instrumentation, Naval Oceanography Command, Bay St. Louis, Mo., 1980.
25. **Jones, M. E.,** *Deepwater Oil Production and Manned Underwater Structures*, Graham & Trotman, London, 1981.
26. **Talkington, H. R.,** Undersea Work Systems, U.S. Naval Ocean Systems Center, San Diego, Calif., 1981.

Chapter 6

MAN VS. THE SEA AT THE SHORE

Orrin H. Pilkey and William J. Neal

TABLE OF CONTENTS

I. INTRODUCTION

In the beginning let us assume that all means of stabilizing open ocean shoreline, i.e., preventing them from eroding, have an adverse effect on the natural shoreline system. This is not to say that therefore no shorelines should be stabilized; that we should allow our major coastal cities to fall in as the sea level rises, or that we should not replace any recreational beaches. The point to be made is that coastal geologists are in complete agreement that our many attempts at halting shoreline erosion have adversly affected the natural system in a variety of ways. The most obvious adverse effect and the one with the greatest sociological impact, has been very significant long-term degradation of our recreational beaches (Figure 1). Less obvious adverse effects include alteration of, or complete removal of natural storm response mechanisms of beaches and natural island migration mechanisms of barrier islands.

In previous years much shoreline stabilization has been carried out as an emergency response to the fact that a house was about to fall in. Situations that are perceived to be emergencies understandably cause little concern for the long range impact on the natural system. In fact, until recently the long range adverse impact of stabilization on the shoreline was not fully understood. In recent years the negative impact of stabilization has become very apparent even if few studies have been made documenting and quantifying stabilization-related changes. This knowledge is slowly beginning to creep into the political system.

There are a number of ways that man interacts with the shoreline. It is critical to recognize these various types of interaction if meaningful and successful coastal planning is to be carried out. We are particularly concerned, in the context of this article, with adverse interaction. One approach to classifying man's adverse interactions is based on the "conditions of activation". We can recognize four major types of adverse interaction between man and the shoreline. These four are (1) fairweather activation, (2) storm activation, (3) long distance activation; and (4) political activation. Before pursuing these types of interaction between man and the shoreline, it is useful to consider the record of the sea level rise in recent years. Because of the expected continuation and acceleration of the sea level rise, increasing conflict between man and the "erosion" problem can be expected.

II. THE SEA LEVEL RISE

Tide gauge records from this century indicate that the sea level is rising (Figure 2). The rate and the cause are not agreed upon; however, the situation is viewed with sufficient alarm that both the National Academy of Sciences and the President's Council on Environmental Quality[2] have pointed this problem out to the American Congress. The rise is sufficiently rapid at this point in time that it may well be the principal cause of world-wide shoreline erosion. Almost all of the world's shorelines are eroding with the notable exception of some deltaic coasts and tectonically rising coasts.

The evidence of sea level rise is mostly from tide gauge records in harbors.[3-5] Such records can be difficult to interpret and many factors may be responsible for apparent sea level changes, (including harbor alterations), but it is generally agreed that beginning in 1930 or so, a more or less continuous rise in sea level has been experienced. The rate may be of the order of 10 to 14 in/century, and some evidence indicates that the rise has accelerated considerably in the 1970s.[5] In January of 1982, two government (NOAA) scientists, Robert Elkins and Edward Epstein, in a report widely quoted by the press, noted that 20,000 m^3 of polar ice may have melted in the last 40 years, contributing significantly to the sea level rise. It is now widely believed that global heating of the atmosphere due to an increase in atmospheric carbon dioxide[6] from the burning of fossil fuel may be at the root of the problem (the greenhouse effect). Particular attention has been focused on the west Antarctic ice sheet which, because it extends well out into the continental shelf, is particularly vulnerable to melting or breaking up.

FIGURE 1. A pocket beach in front of the seawall at Cape May, N.J. At high tide no beach exists here at all. Pocket beach swimming at low tide is becoming more and more common on American beaches.

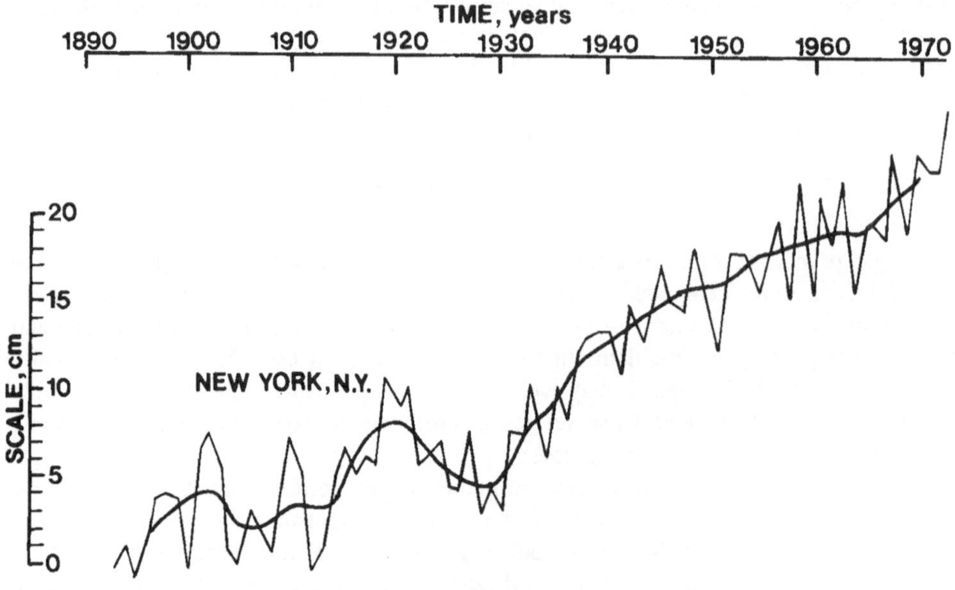

FIGURE 2. The sea level rise in New York harbor. (From Hicks, S. D., *Shore and Beach*, 40(1), 20, 1972.)

The relationship between sea level rise and shoreline retreat is a simple one when viewed from time spans of hundreds or thousands of years.[7,8]

That is, the shoreline must retreat rapidly in a horizontal direction over flat surfaces such as a coastal plain and must retreat much less rapidly on steeper mountainous coasts (Figure

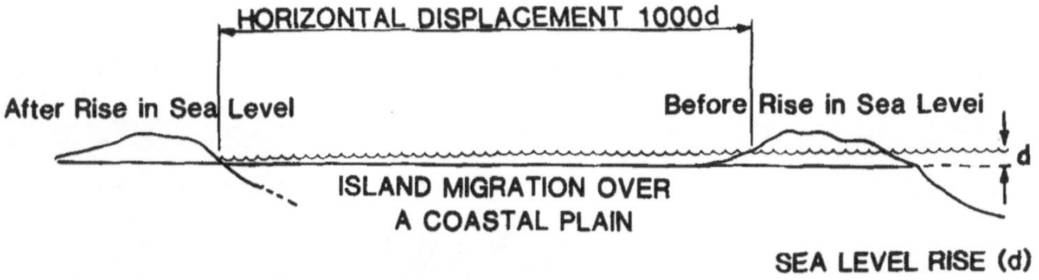

FIGURE 3. The relationsip between the sea level rise and horizontal shoreline retreat on a coastal plain coast. A small rise produces a large horizontal retreat.

3). The ratio of rise to retreat over much of the American coastal plain should be of the order of between 1 to 500 to 1 to 1500 and more. This is a simple geometrical truth.

The problem, when viewed in a shorter time frame, is more complex. The barrier island shorelines are not retreating as rapidly as they should be if geometry of the flooded surface was the only control. We should expect retreat of the open ocean barrier shorelines to be of the order of 10 ft/year instead of the more normal 3 ft/year we usually observe. Thus, there is a starting up problem. Swift's[9,10] model of coastal sedimentation in a rising sea level must apply to this problem, but again, his studies apply to too long a time period to be useful for man's immediate need-to-know, from a coastal development perspective.

Predicting what the short-term response to rising sea level will be is an area of critical research need. It is well known that shorelines are in a dynamic short-time equilibrium involving sand supply, wave energy, beach shape, as well as the rise of sea level. In addition, slope of the shoreface, slope of the shoreline migration surface, stratigraphy of the shoreface, and a number of other factors must play a role in shoreline migration or retreat rate.

III. MAN VS. THE SHORELINE

A. Fairweather Activation

Fairweather adverse interaction of man on the shoreline refers to situations where structures and objects built by man interface with the dynamics of the shoreline during typical or normal fairweather conditions. Thus any structure built or emplaced between the low and high tide lines is one that must fall in this category. By far the most common means of such interaction are groins and seawalls (Figure 4).

The seawall effect is well documented by the U.S. Army Corps of Engineers.[11] It is fair to say that in almost all circumstances, the effect of a seawall will be to severely degrade the recreational beach. Seawalls reflect wave energy and intensify longshore currents. In either case, the long range response is a greatly steepened shoreface and a narrowed beach (see Figure 4).

Seawalls cause other problems as well. They prevent storm wave access to dunes which supply sand temporarily to the lower beach in order to flatten the system and/or build offshore bars. On a flattened beach wave energy is dissipated over a broadened zone. Offshore bars, in effect, trip the larger waves before they reach the beach.[12] Either type of beach storm response is prevented with a seawall blocking access to the requisite sand supply.

In a longer-range and perhaps more fundamental sense, seawalls cause problems because they hold the shoreline in place as the sea level rises and the equilibrium shoreline position moves landward.

Groins are a large family of engineering structures consisting of some sort of wall built perpendicular to the beach for the purpose of broadening the beach by trapping sand moving

FIGURE 4. The sea wall at Sea Island, Georgia. The wall was built in response to Hurricane David which brushed by the island in 1978. Now no beach exists at high tide and rubble from the wall is beginning to make the low tide beach difficult to use.

in the longshore transport system. The problem is that the sand was moving somewhere, and where ever somewhere is, it now has a sand supply problem. In order to alleviate the problem, a variety of ingenious approaches have been used that will allow sand to flow through or over the groin, but these never successfully replace the natural sand supply system.

Stabilized shorelines with fairweather activation are by far the most susceptible to rapid deterioration of the natural shoreline system (Figure 5).

B. Storm Activation

Certain types of coastal stabilization structures, in particular bulkheads and revetments, are basically supposed to work only in time of storm activity. That is, they are built above the normal high tide line. There is little disagreement among coastal geologists and engineers that fairweather activated structures cause erosion or beach narrowing in the long run. There is, however, still some disagreement as to the extent to which revetments cause erosion. Revetments are usually made by covering the face of the first dune with boulders. The principal involved here is that during storms the wave swash striking the boulders will be temporarily absorbed in whole or part by the interstices between the boulders. This should reduce backwash which is believed to be a major cause of beach narrowing. Also, the presence of the boulders is a direct deterrent to dune erosion.

Bulkheads are best described as seawalls built above the high tide line. They are a wall built parallel to the shoreline and expected to come into contact with the surf zone only in the case of extraordinary tides or storms.

Both bulkheads and revetments must cause erosion in two ways. First, as in the aforementioned example of the seawall, they prevent the natural storm response (Figures 6 and

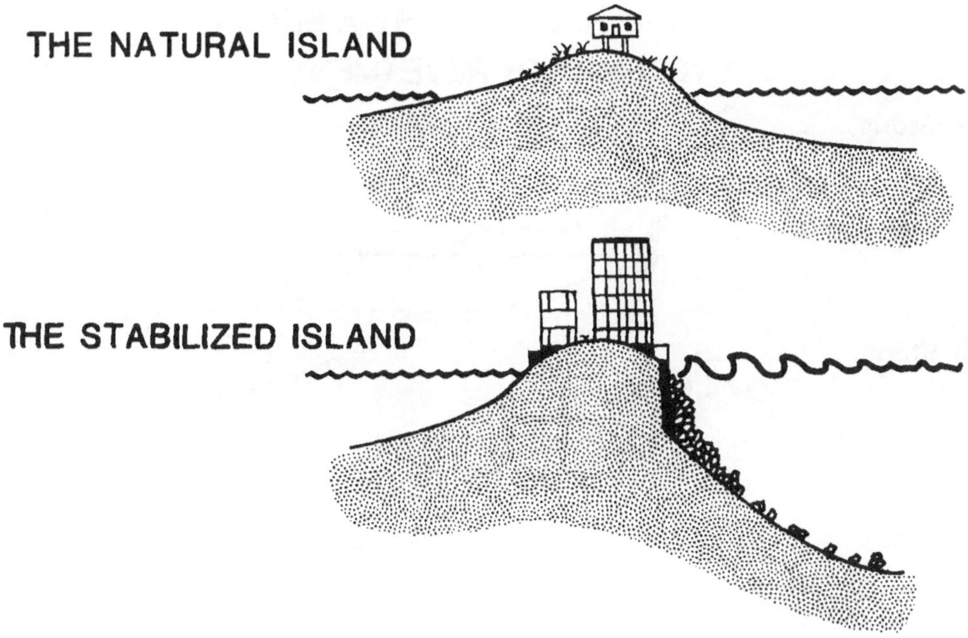

FIGURE 5. A diagrammatic illustration showing the long range (10 to 50 year) effect of stabilization on a shoreline system. In the final stage the armored shoreline sits atop a steepened shoreface and all waves strike the seawall.

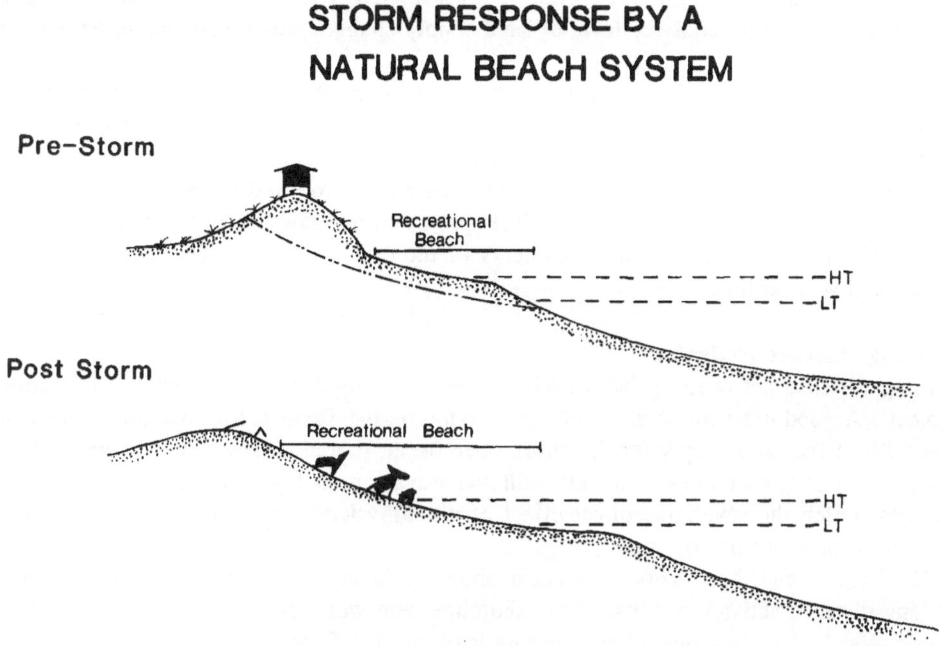

FIGURE 6. The storm response at a natural beach system. The flattened beach is capable of dissipating wave energy over a broadened surface area relative to the pre-storm beach.

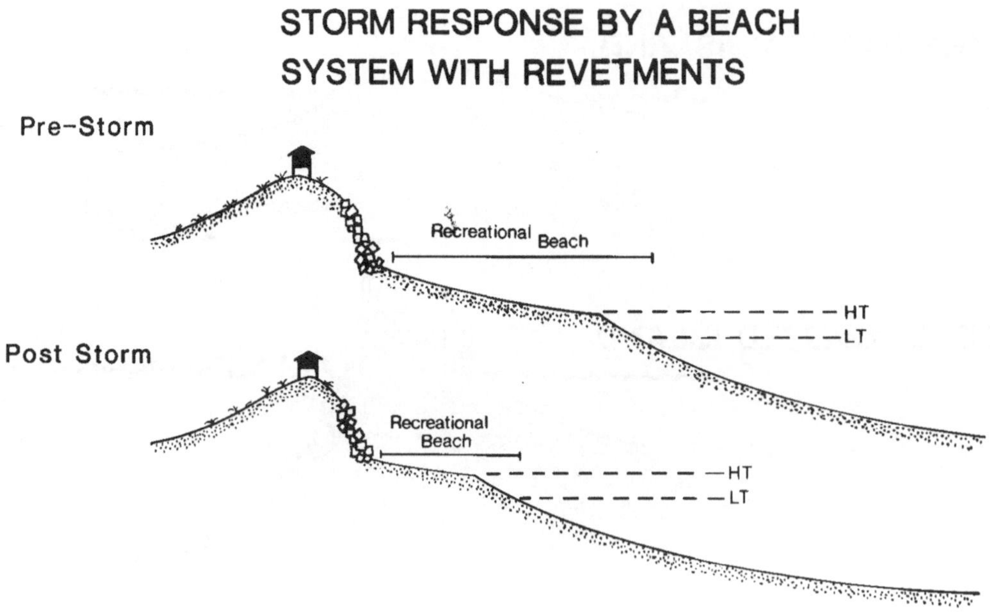

FIGURE 7. The revetted beach storm response is a steepened shoreface. Compare this with Figure 6. On the plus side the revetment kept the house from falling in.

7) from taking place. The beach, instead of flattening or forming offshore bars may suffer irreversible sand loss in storms. Second, bulkheads and revetments hold the shoreline in a fixed position regardless of sea level or sand supply changes which may cause the natural beach to move landward (i.e., erode).

In general, storm activated structures cause damage to the shoreline system more gradually than fairweather activated systems. However, because storm activated systems do cause erosion in a longer-range sense, they must, by definition, eventually become fairweather activated systems. Putting it in another way, shorelines protected by bulkheads and low revetments will eventually become shorelines protected by seawalls. With shoreface steepening and the resulting increased wave energy on the wall, all seawalls become grander and more massive with time.

C. Long Distance Activation

Long distance activation refers to adverse effects caused by "upstream" stabilization projects. A good example of this problem is found on the Texas coast. Morton[13] estimated that 50% of the sand supply for the entire open ocean shoreline is tied up by jetties. This must mean that erosion rates, statewide, will increase and probably even accelerate in coming decades. Often, however, the direct effect of a nearby jetty, groin, or seawall system is difficult to prove or demonstrate.

The Long Island, New York open ocean shoreline is another example of adverse effects of long distance activation. Most of the shoreline here was eroding about 1 ft/year before World War II. Now the overall erosion rate is closer to 6 ft/year.[14]

The difference is quite likely due to a large number of groins, revetments, jetties, seawalls, and other forms of stabilization. Essentially, unstabilized portions of the shoreline, e.g., Fire Island, New York, are as much affected as more heavily developed islands.

The east-west New York barrier shoreline is supplied by sand from bluffs of glacial material eroding at the east-end of the island chain. As development pressure continues,

FIGURE 8. An undeveloped shoreline in Georgia that is eroding rapidly. Photo by Bob Frey.

more and more of this shoreline will be stabilized by seawalls and revetments. Eroding bluffs will be stabilized to save homes, e.g., at Easthampton, New York, but the "downstream" beaches will starve. Quite frequently on our Pacific and New England coasts, bluffs and cliffs will be reinforced to prevent failure or erosion with no thought at all for the resulting long distance activation of detrimental effect on the shoreline miles away.

D. Political Activation

The interaction of man and the shoreline cannot be viewed totally in an engineering or a geological context. Community decisions as to how to respond to an eroding shoreline (Figure 8) are usually highly political in nature. Of course, community, state, and federal economics play a role too, but political considerations often supersede all other considerations, including scientific.

Adverse shoreline effects that could be termed politically activated could include just about any type of shoreline stabilization project. However, in some cases, the problem is more purely political.

Decisions to build or enlarge port facilities generally set off a chain of events leading through several possible pathways to shoreline stabilization. For example, establishment of a new "fish park" at Wanchese, North Carolina has led to strong support and detailed planning for a large jetty system to protect nearby Oregon Inlet. Deepening of inlet channels for harbors, in general, may eventually lead to jetty emplacement and hence stabilization of the shoreline. The deeper the channel the more costly to maintain by dredging in the dynamic nearshore zone. Jetties in turn lead to active interruption of the shoreline sand supply system and hence increased rates of erosion. Increased rates of erosion may lead to fairweather activated or storm activated stabilization measures. Even if no jetty system is built, the relatively simple act of digging a deeper channel will usually cause disruption to the sand system which then snowballs into other forms of stabilization.

FIGURE 9. A motel on Topsail Island threatened by erosion, November, 1982. Note the exposed septic tanks on the beach.

A major problem in shoreline planning has been ignoring the snowball effect. Deepened channels are evaluated as deepened channels, that is, their effect is considered only from the viewpoint of short distance and short time responses.

Another example of political activation of shoreline problems is illustrated by beach replenishment. Frequently when beaches are replenished for the first time, a beach community virtually comes alive (Figure 9). The beach scene changes from the depressing situation where houses and motels are threatened or have actually fallen in to a broad, smooth beach just like the old days. Property values jump and single unit housing is replaced by multiple unit and sometimes high rise construction. Replenished beaches erode more rapidly than natural beaches because only the upper foreshore is replaced (Figure 10). If one considers the entire beach system to a depth of 10 m or so, the replenished sand can be viewed as a pile of sand at the top of a long ramp. The replenished sand oversteepens the beach profile which comes back to its natural shape relatively rapidly. Thus, when the next replenishment is needed, the community has a much larger political voice because the first replenishment project gave the community time to build up.

Replenishment cannot be carried on indefinitely. It cost $65 million to replenish 15 mi of Miami Beach. Generally beach replenishment costs a minimum of $1 million/mi. At the same time along any developed shoreline long distance activation will decrease the sand supply further. Eventually more "severe" structural measures of stabilization can be expected to come into play. In this fashion, political activation leads in the long range (10 to 50 years) to fairweather activation.

IV. OREGON INLET

A. Anatomy of a Coastal Crisis

The controversy swirling around the proposed jetties at Oregon Inlet on North Carolina's

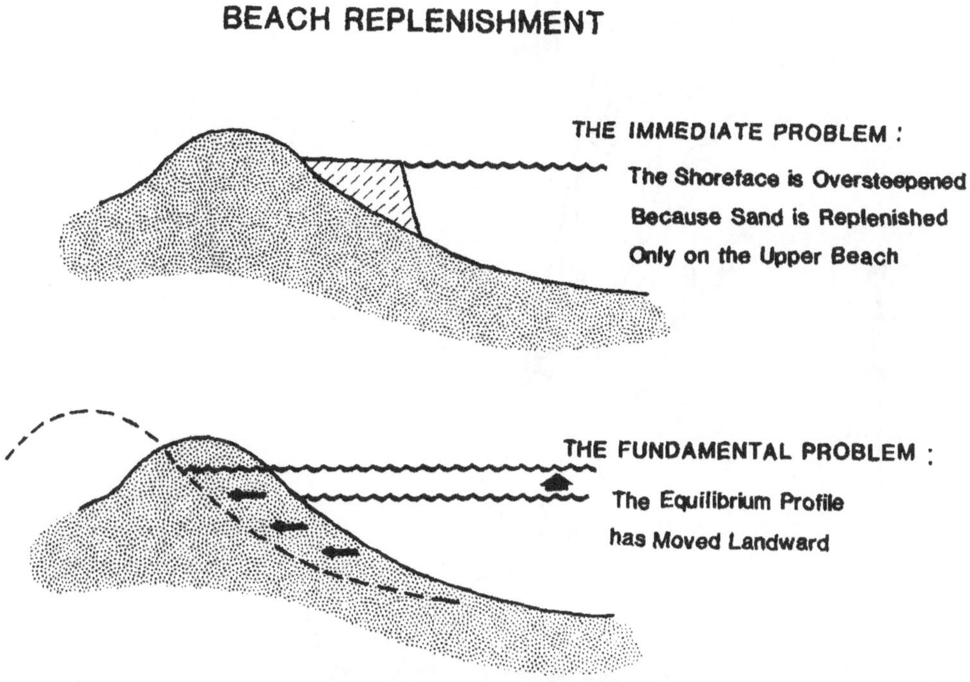

FIGURE 10. A diagrammatic illustration showing why a replenished beach disappears more rapidly than its natural predecessor. The first problem is that the sand is piled atop the upper beach which steepens the whole system. In order to get back to a profile in equilibrium with local sand supply and wave climate conditions, the sand rapidly disappears across the entire shoreface.

Outer Banks (Figure 11) is an example that encompasses all of the ingredients of environmental clashes at the shoreline. The ingredients include state and local politics, fishing economics, national seashore and wildlife preserve management, and the engineering mentality vs. the scientific mentality. It is also an example of political activation of shoreline stabilization with all of the potential long range problems caused by stabilization.

Oregon Inlet formed during a hurricane in 1843. It very likely formed, as most other Outer Banks inlets have formed, as a result of water rushing seaward from the lagoon (Pamlico Sound) as the storm passed inland. After the inlet formed it began to be used by local fisherman on Roanoke Island, particularly from the small village of Wanchese. The State of North Carolina put up a small lighthouse on the south side of the inlet which quickly fell in as the inlet migrated from north to south, the dominant direction of longshore drift on the barrier islands. A second lighthouse met with the same fate, so the Bodie Island light, which still exists, was built on the north side of the inlet. In a few decades the inlet had moved so far in a southerly direction that the light became essentially useless. It now sits more than a mile north of the inlet it was supposed to guide fisherman through. The trend for more than 100 years has been a steady migration of the inlet at a rate of about 75 ft/year.

In the late 1950s the State of North Carolina began to plan construction of a bridge across Oregon Inlet. The Corps of Engineers repeatedly warned North Carolina state officials that the bridge should not be built because of the inlet's historical migration pattern. Nevertheless, construction began in 1961. In 1962, during the Ash Wednesday storm, the inlet widened in a period of less than 14 hr from the normal 1/2 mi width to 2 mi (Figure 12). The inlet had basically "blown its cork" by acting as an outlet for huge volumes of water pushed into Pamlico Sound by this 3-day northeaster that struck at spring tides.

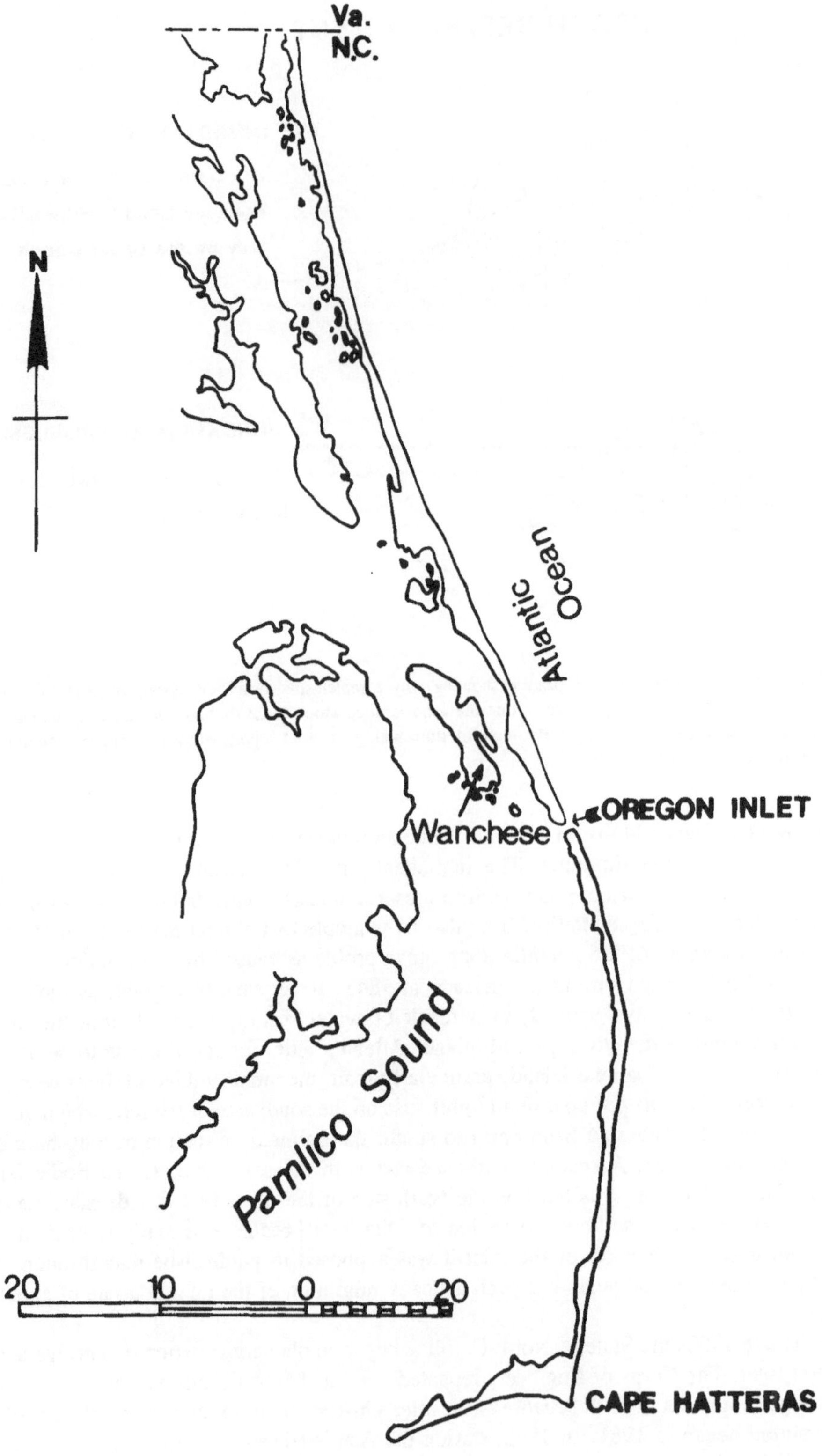

FIGURE 11. Location map showing Oregon Inlet on North Carolina's Outer Banks.

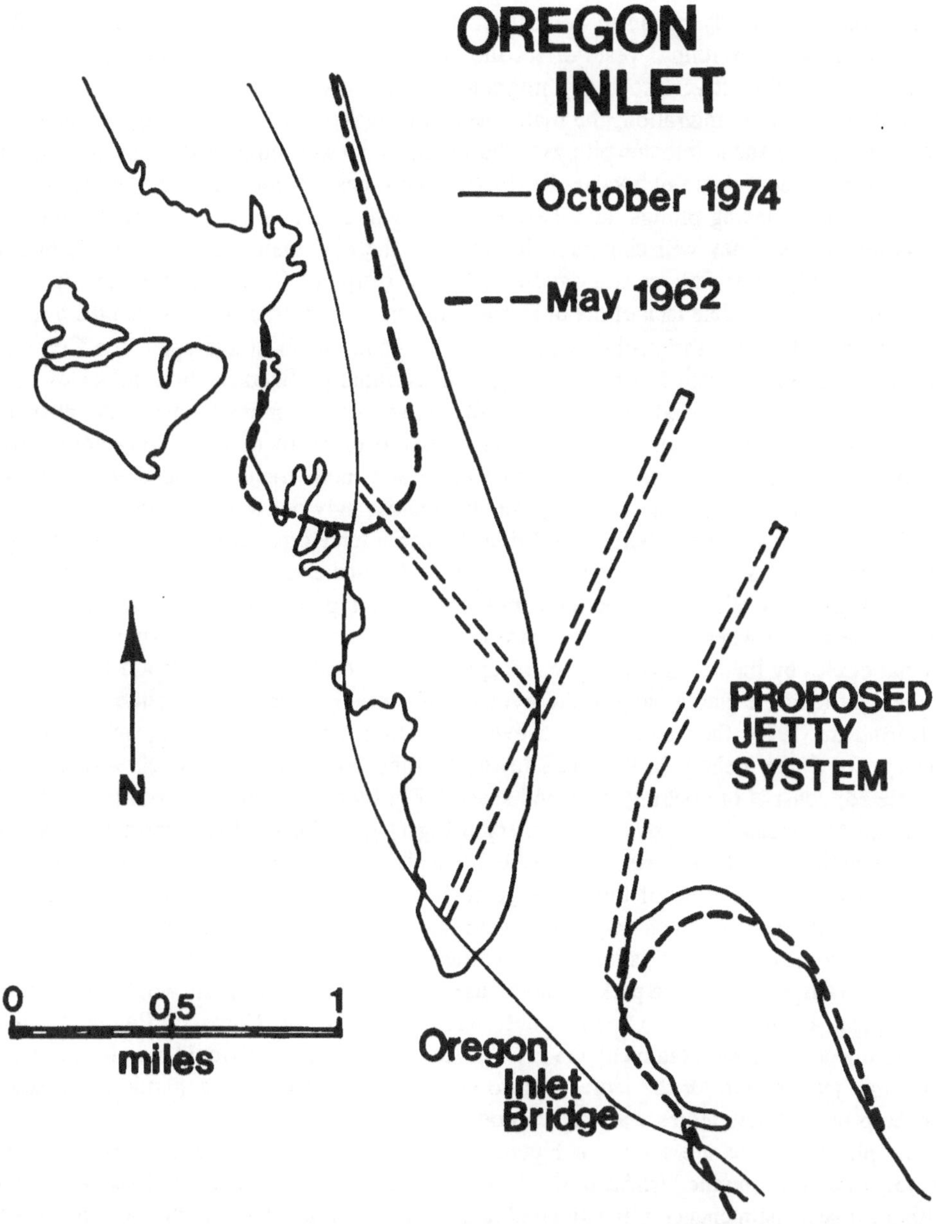

FIGURE 12. Oregon Inlet, showing the configuration of the proposed jetties and the location of the inlet boundaries immediately after the 1962 Ash Wednesday storm.

The pilings and abutments already emplaced on the north side of the inlet disappeared overnight, along with some heavy construction equipment. An examination of available navigation charts and areal photos indicates that abrupt widening and deepening of Oregon Inlet has occurred previously as a result of storm activity. This is caused by the fact that Pamlico Sound is a large body of water capable of temporarily storing very large volumes of storm surge water. Inlet blowout, i.e., major widening, is to be expected in any major storm which forces water into the sound.

Undaunted, the State began construction of the bridge anew. By the time it was completed, but before the first tourist drove across it, fishing platforms at the north-end were over dry land as the inlet slowly began to come back to the 1/2 mi width required to handle the normal tidal prism. Within 5 years or so, the inlet was once again a 1/2 mi wide, and the normal pattern of 75 ft/year southerly migration began once again.

As a result of inlet migration, the main channel began to move out from under the main span of the bridge and to threaten pilings to the south, which were emplaced on the assumption that such migration would not occur. By 1980, three times the initial cost of the bridge had been spent in replacing pilings, and these expenses will continue and accelerate as long as the bridge exists. They will continue, that is, until the next major storm when the bridge almost certainly must fail, as was predicted by the Corps of Engineers prior to the 1962 storm. Only the complete lack of major storms since it was constructed has saved the bridge.

Meanwhile, because shrimp boats must pass under the main bridge span, the Corps is forced to maintain the main channel by dredging in essentially a fixed position. Inlet closures, the latest of which occurred in winter of 1982, have become more frequent. As channel maintenance continues against the continual pressure of north to south longshore drift, closures during minor storms must become more and more frequent and dredging costs should rise in proportion. The inlet will become increasingly out of equilibrium with time.

What alternatives existed before the bridge? Once a ferry crossed the inlet, but a bridge was deemed to be a more economical approach. Unfortunately, the economic evaluation or comparison did not take into account the inevitable inlet migration and the equally inevitable loss of the bridge when the "right" storm comes by. Furthermore, the distress caused in official circles by the accelerating costs of dredging and bridge salvation led to suggestions of the next (and inevitable) step in the pyramid of shoreline stabilization; jetties.

During the 1970s the State received federal funding for a seafood park, an innovative concept for a new fishery in the small town of Wanchese. The park, which was recently completed, consists of docking areas and space for onshore fish handling facilities and was supposed to increase the size of the fishery using Oregon Inlet. Most important of all, it was hoped that it would increase employment in a chronically depressed section of the state. Unfortunately, at the time of this writing, no fishing customers have been found for the park. The new facility stands empty and waiting, although local fish landing has increased significantly using the pre-existing dock facilities. Local marine scientists and environmentalists who suspected that the park would be used to justify jetties (that is, would be a source of political activation), for Oregon Inlet were assured in the Corps of Engineers final Environmental Impact Statement (1977) that the seafood park "would not be used for economic justification for the jetties." Two years later the principal economic justification for the proposed new jetties was the seafood park.

The planned jetties are shown in Figure 12. They are unusual in several respects most notably in their economic justification. The cost of the jetties is estimated to be $100 million with a 50 year maintenance bill estimated at $500 to $600 million. Yet the present fishery to be supported by the jetty project is very small, a few dozen boats. The economic justification is therefore based on a new "trash fishery," i.e., it is assumed that American fishermen will begin to catch the so-called "trash fish" formerly caught by Russian and East German trawlers before imposition of the 200 mi limit. The trash fish are called this because although they are edible (for example squid and hake), they have relatively little value in the American market. U.S. Fish and Wildlife Service officials consider the economic justification contrived. Although no one can deny that such a fishery is a possibility (1) there is no indication of one starting up anywhere in the U.S., and (2) if a "trash fishery" were to begin it would be much more likely to start in the northeastern U.S. Foreign fleet catch records indicate that the catch in northern waters was 50 times that of the catch in Virginia or North Carolina waters. Why would a business take a chance in starting a fishing industry in the south?

The proposed jetties are on the Cape Hatteras National Seashore and a short distance south of the proposed jetties in a downstream direction, the Fish and Wildlife Service maintains large duck and geese nesting ponds. Both the National Park Service and the U.S. Fish and Wildlife Service have strongly opposed the jetties because of possible increased erosion rates on Pea Island refuge to the south.

The National Park Service empaneled a group of five shoreline scientists (consisting of one engineer, one biologist, and three geologists) headed up by Dr. Douglas Inman of Scripps Institution of Oceanography. The panel wrote a report strongly opposing the jetty plan. Some of the reasons for their opposition to the jetties were:

1. The Corps proposed to bypass sand from north to south every one or two years. The panel felt that sand movement is erratic and particularly large volumes may move in a single storm. The artificial sand bypass system would not in any way be related to the timing of sand movement in the natural system and as a consequence considerable erosion might occur.
2. The sand bypass system depends on a floating breakwater system necessitated by the very high wave energy regime of the Outer Banks. The breakwater is to be hauled in and out of the inlet when dredging is to be carried out. The panel noted that the breakwater design was not complete and possibly may not even be feasible within reasonable economic constraints. Nonetheless, the Corps put a price tag on the yet-to-be-designed breakwater for the purpose of calculating cost-benefit ratios.
3. One of the suggested benefits of the jetties is stabilization of an eroding barrier island. No single aspect of the Corp's design of the jetties more clearly illustrates the difference between the engineering mentality and the scientific mentality. The Outer Banks of North Carolina are generally eroding on both the front side and the backside. The geologic view is that this is the expected response to a rising sea level. The islands will narrow to the point that overwash occurs readily during storms and then genuine landward migration will occur. The Corps of Engineers official opinion of the Outer Bank erosion phenomenon is that island narrowing is not part of the island migration phenomenon. The Corps view is that the islands are in danger of disappearing. The panel view was that the islands are in the process of migrating. If the islands are about to disappear, which is the Corps view, then perhaps the jetties could serve a useful stabilization role. Modern geologic evidence (Figure 13) is much to the contrary; the islands are in no danger of loss, only in danger of changing their shape and position.
4. The jetties were designed on the assumption that the substrate on the shoreface was sand. Jetties can be constructed on any type of marine substrate, but different designs are required for different substrates. The problem was that standard (and out of date) dogma dictates that barrier island shorefaces are sand. We now know that significant thicknesses of mud may be present; mud that has passed from the lagoon side to the open ocean side as the island migrated.[9] Such mud is an inevitable by-product of island migration by overwash. Possibly as a result of pressure from the state marine science community the Corps drilled and cored the shoreface in suitable fashion for the first time in the Fall of 1981. Significant thicknesses of mud were found, necessitating design changes long after the environmental impact statements, design memoranda, and cost-benefit ratios (1:1.14) were in the public record.
5. The Corps of Engineers made a sophisticated and costly model study of Oregon Inlet at their model facilities in Vicksburg, Mississippi. The model purports to show that the jetties will not cause flooding or a blowout elsewhere during the next storm. The problem with model studies is that they are only as good as the quality of the oceanographic and other data that are used in the model. One of the objections of the National Park Service panel of scientists was that they could not determine exactly what field data were available to use in the model, but it appeared that only a small amount of real world data had been obtained.

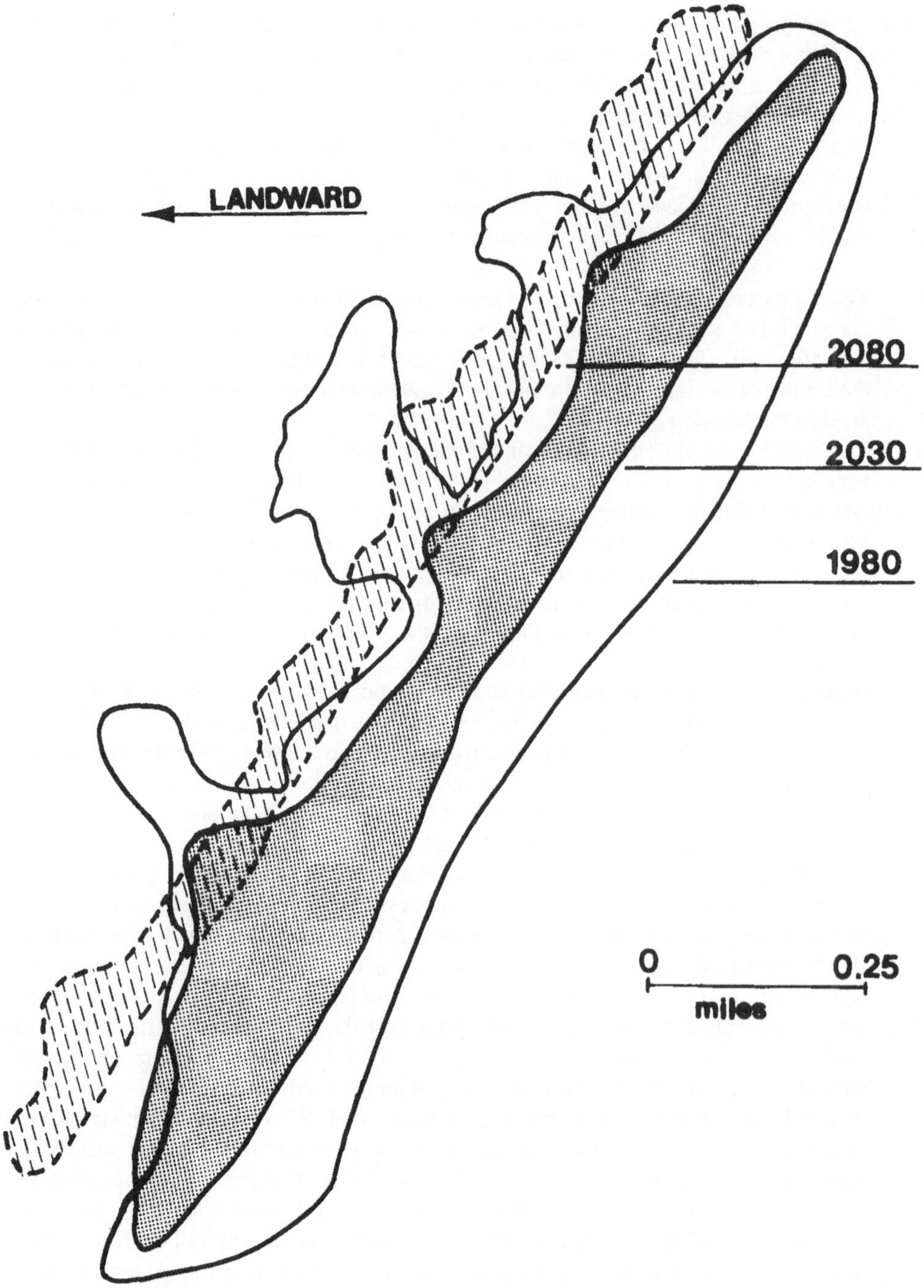

FIGURE 13. A diagrammatic view of the probable future of North Carolina's Outer Banks. At the present time the islands are narrowing. After they become narrow enough, overwash will readily cross over the islands at a high enough frequency to allow the island to migrate. The islands migrate through front side erosion and backside widening.

In any event, such models have proven in the past to be poor at estimating the long range (10 to 50 years) effect of jetties on open ocean shorelines. Part of the problem is that jetties may cause major changes in shape of the adjacent shoreface, causing changes in wave fraction patterns with attendent changes in sand transport quantities and direction, beach storm response, and many other parameters.

The use of models is another example of the engineer-scientist conflict at the shore. Geologists tend to be skeptical of their value in predicting success or failure of jetties in any meaningful time frame. On the other hand, models by their very cost and sophistication, tend to silence opposition to coastal projects because they are difficult, if not impossible, to respond to formally without a great deal of technical effort.

The Oregon Inlet jetty project is presently stalled in the Congress, probably a casualty of the present economic climate. The project is strongly backed by the state government and especially local commercial fishermen, who daily traverse this very dangerous inlet. It is strongly opposed by the National Park Service, the National Fish and Wildlife Service, environmental groups, and the North Carolina Marine Science community. This Corps project has probably received more exposure to the "outside" scientific community that any other similar project in recent decades. Non-Corps scientific opinion is essentially unanimous that the project is poorly conceived in terms of design assumptions. However, all of this seems to have had little impact on the jetty planning. The Oregon Inlet jetty system is a classic example of the difficulty of getting sound scientific advice into the political arena of the shoreline.

V. SUMMARY

In 1981, a position paper was issued by the coastal geology community. It was signed by 110 individuals, the vast majority of the research scientists in this speciality. The summary of the position paper reproduced below essentially summarizes many of the points made in this paper and also briefly addresses the solutions to the problem.

A. The Position Paper Summary

New approaches to the management of the American shoreline are urgently needed to preserve our recreational beaches for future generations. Approximately half of the 10,000 mi of the "lower 48" American shoreline facing the open ocean is under development pressure. Well over 2,000 mi are considered by the U.S. Army Corps of Engineers to be in a state of critical erosion. Erosion is occurring along almost all of the U.S. coast and when shoreline retreat collides with shoreline development, a state of "critical erosion" is achieved. Shoreline retreat is due to many causes but a major one is rising sea level and indications are that the rise will continue for the foreseeable future.

The usual response to critical erosion on America's shore is stabilization; halting of shoreline retreat by engineering means. Such stabilization of America's shore has been successful in increasing the length of life of buildings built adjacent to the beach. However, stabilization in the long run (50 years ±) and sometimes in a much shorter time frame has resulted in severe degradation of the recreational beach area. Dollar costs of halting shoreline retreat by stabilization is very high. Replacement of the beach by pumping in new sand costs about $1 million or more per shoreline mile each time it is done and it must be carried out repeatedly; commonly in 3 to 10 year intervals. Another approach, the building of seawalls, costs between $100 to $600 per linear open ocean shoreline foot. Combining these costs figures with the 2,000 mi figure of critically eroding shoreline gives some idea of the magnitude of the potential economic crisis on the American shoreline if we continue to stabilize.

American taxpayers are paying huge sums of money to temporarily protect the private property of a relative few. Furthermore this practice commonly leads to the ultimate destruction of a highly valued public recreational area.

Stabilization costs can be justified for major coastal cities or harbor entrances (Chicago,

Galveston, Miami Beach, Coney Island, the Columbia River entrance, for example), but stabilization of most American shores is not justifiable in the broader scope of national interests. Numerous projects, involving public and private money along virtually all developed coastal and lake shores presently threaten most of America's recreational shoreline.

The following points summarize our views on stabilization of America's open ocean shorelines:

1. People are directly responsible for the "erosion problem" by constructing buildings near the beach. For practical purposes, there is no erosion problem where there are no buildings or farms.
2. Fixed shoreline structures (breakwaters, groins, seawalls, etc.) can be successful in prolonging the life of beach buildings. However, they almost always accelerate the natural rate of beach erosion. Resulting degradation of the beach may occur in the immediate vicinity of structures or it may occur along adjacent shorelines sometimes miles away.
3. Most shoreline stabilization projects protect property, not beaches. The protected property belongs to a few individuals relative to the number of Americans who use beaches. If left alone, beaches will always be present, even if they are moving landward.
4. The cost of saving beach property by stabilization is very high. Often it is greater than the value of the property to be saved especially if long range costs are considered.
5. Shoreline stabilization in the long run (10 to 100 years) usually results in severe degradation or total loss of a valuable natural resource, the open ocean beach.
6. Historical data show that shoreline stabilization is irreversible. Once a beach has been stabilized, it will almost always remain in a stabilized state at increasing cost to the taxpayer.

The consequences of responding to rising sea level by shoreline stabilization are so serious that we urge immediate measures to explore totally new approaches to shoreline management. Such approaches may even involve drastic and unpopular measures such as assuming that buildings adjacent to the beach are temporary or expendable. Equally important, the new approach to shoreline management must incorporate the very significant advances in geologic understanding of shoreline processes that have occurred during the last decades. In the past the American public has been largely unappraised and unaware of the long range environmental and dollar costs of shoreline stabilization. There is a critical and immediate need for the public to know the direction in which American shoreline mangement is leading.

REFERENCES

1. **Hicks, S. D.,** Long-period sea level variations for the United States through 1978, *Shore and Beach,* 49(2), 26, 1981.
2. **Marshall, E.,** By flood, if not by fire, CEQ says, *Science,* 221, 463, 1980.
3. **Hicks, S. D.,** On the classification and trends of long period sea level series, *Shore and Beach,* 40(1), 20, 1972.
4. **Hicks, S. D. and Jones, E. C.,** Trends and variability of yearly mean sea level, NOAA Tech. Memo #13, 39, 1974.
5. **Emery, K. O.,** Relative sea levels from tide-gauge records, *Nat. Acad. Sci., U.S.A.,* 77(12), 6968, 1980.
6. **Hansen, J., Johnson, D., Lacis, A., Lebedeff, S., Lee, P., Rind, D., and Russell, G.,** Climate impact of increasing carbon dioxide, *Science,* 213, 957, 1981.

7. **Pilkey, O. H., Jr., W. J. Neal, and O. H. Pilkey, Sr.,** From Currituck to Calabash: living with North Carolina's barrier islands, North Carolina Science and Technology Research Center, Research Triangle Park, N.C., 1978, 228.
8. **Leatherman, S. P.,** Barrier beach development: a perspective on the problem, *Shore and Beach*, 49(2), 2, 1981.
9. **Swift, D. J. P.,** Barrier-island genesis: evidence from the central Atlantic shelf eastern U.S.A., *Sed. Geology*, 14(1), 1, 1975.
10. **Swift, D. J. P.,** Coastal sedimentation, in *Marine Sediment Transport and Environmental Management*, Stanley, D. J. and Swift, D. J. P., Eds., John Wiley & Sons, N.Y., 1976, 602.
11. **Corps of Engineers,** Shore Protection Manual Vol. I, II, and III, 2nd Ed., U.S. Government Printing Office, 1975.
12. **Hayes, M. O. and Boothroyd, J.,** Storms as modifiying agents in the coastal environment, in *Coastal Environments: Northeast Massachusetts*, Department of Geology, University of Massachusetts, Amherst, 1969, 290.
13. **Morton, R. A.,** Temporal and spatial in shoreline changes and their implications, examples from the Texas Gulf Coast, *J. Sed. Petrology*, 49, 1101, 1979.
14. **McCormack, L.,** personal communication, 1982.

Chapter 7

COASTAL MANAGEMENT: AN UNFINISHED UNDERTAKING

Allen Miller

TABLE OF CONTENTS

I. INTRODUCTION

"The Congress finds and declares that it is the national policy (a) to preserve, protect, develop, and where possible, to restore or enhance, the resources of the nation's coastal zone for this and succeeding generations. . . ."[1]

With this grand statement of policy the 92nd Congress, on October 23, 1972, launched the federal government's efforts to improve management of the nation's oceanic and Great Lakes' coasts. Thwarted by the initial impoundment of funds, the ambiguous language of the Act, opposition from the American Petroleum Institute and other special interest groups, a relatively small budget and a lack of precedences, Coastal Management nonetheless continued on course to improve state comprehensive management programs for the nation's coast. By October of 1980, 25 states had met federal criteria for an approvable program.[2] The 95th Congress, impressed by both the progress and the success of the program, reauthorized the effort for five additional years, only to discover in January of 1981 that the program had largely "achieved its purpose".[3]

This chapter is a review of the history of the Coastal Zone Management Act and of selected state coastal management authorities, the numerous and often critical assessments of the national program, amendments to the Act, and a look to the future.

II. EVOLUTION

A. Federal Legislation — Towards Comprehensiveness

Much has been written on the history of the federal Coastal Zone Management Act of 1972. Prior to its enactment federal ocean concerns had been expressed through functional programs addressing specific issues such as recreational access, fisheries management, or wetland protection. It was not until the Coastal Zone Management Act that a multitude of coastal concerns were to be addressed as one system under a coordinated management effort.

In the article entitled A Legislative-Political History of the Coastal Zone Management Act of 1972, Zigurds L. Zile provides an enlightened description of federal involvement in Coastal Management. Zile describes the Act as evolving through four phases of development: recreation, estuary protection, ocean development, and land use policy.[4]

1. Recreation Phase

The recreation phase began with a 1934 survey of coastal areas of highest recreational value, and continued through the National Seashore and National Lakeshore Programs of 1937 to the Land and Water Conservation Program of 1964. Despite vast public expenditures, the recreational phase failed to achieve the major impact desired of preservation through public ownership. In 1973, 40 years after the initial survey, the National Water Commission found that of the 1/3 of the 60,000 mi of coast in the lower 48 states that had recreational potential only 1% were publicly owned.[5] Lack of money was, and continues to be, the limiting factor in public acquisition programs.

2. Estuary Protection Phase

By the 1960s, society's postwar interest in leisure activity was being complemented with a rapidly increasing understanding of the natural environment and an interest in resource protection. A "Long Island National Wetland's Recreation Area", similar to the national seashore concept, was proposed in the House of Representatives in 1965.[6] In proposing federal management of this unique area, the bill met with a hostile response from both state and local government over what was termed the "unwanted intrusion of a federal administrative department".[7]

In 1966 Representative Dingell of Michigan expanded the idea of wetland protection to a national system of estuarine areas.[8] Dingell met the same response, resistance by the states who viewed federal involvement as a threat to their prerogatives and authorities. The debate over federal wetland protection continued to the passage of "The Estuary Protection Act" in July, 1968.[9] The Act, however, did not protect wetlands. It called only for the Department of Interior to study the problem and to inventory the nation's estuaries. This report, known as the "National Estuary Study", was completed in January of 1970 with seven volumes of data and few recommendations.[10] Despite the limited recommendations the report stated that, "the Coastal and Great Lakes states are the key to successful estuary protection and restoration."[11]

Concurrent with the Interior study was the "National Estuarine Pollution Study" authorized under the Clean Water Act of 1966.[12] Completed in November of 1968, the report concluded that "management problems of the estuaries were inseparably related to other resources of the coast, and any management system had to deal with the coastal zone in its entirety." The state role was to manage. The federal role was to assist and to coordinate federal agencies.[13]

The message to Congress was becoming clear. Neither federal nor the state government could buy all of the areas which needed protection, nor were state and local governments going to allow the federal government to extend into areas which they viewed as their prerogatives. Most importantly, as stated in the "Estuarine Pollution Study", management of the coastal resource could not be successfully accomplished through separate functional programs. The coast was to be viewed and managed as a unit.

3. Oceans Development Phase

While the interest in recreation access and resource protection continued, others began viewing the coast as an opportunity to make better use of the resources found there. The "Marine Resources and Engineering Development Act of 1966" established the Commission on Marine Science, Engineering and Resources to: (1) "examine the nation's stake in the development, utilization, and preservation of our marine environment"; (2) to review current and contemplative marine activities; (3) to "formulate a comprehensive long-term national program"; and (4) to recommend a government organization to implement the program.[14] The Commission, popularly known as the Stratton Commission, reported to Congress in January of 1969 in *Our Nation and The Sea.*.[15] The Commission report became the primary source for future congressional legislative efforts.

The problem was management, said the Commission, and the burden of managing the coast was on the states.

"Effective management to date has been thwarted by a variety of government jurisdictions involved, the low priority afforded marine matters by state government, the diffusion of responsibility among state agencies, and the failure of state agencies to develop long-range plans."[16]

The commission's view was that a series of state programs would be the national program. States were to voluntarily participate in the federal effort to manage the coast in exchange for the promise of limited federal financial assistance and the authority to require federal agencies to be consistent with their programs. "Even the most ardent federalists amongst us despair of ever doing an effective job at the federal level", testified Dr. John A. Knauss, Chairman of the Panel on Coastal Management. Knauss viewed the federal role as "coordinating federal activities in these areas, and for planning, coordination, and the management of that part of the coastal zone within federal but beyond state jurisdiction."[17]

These views were shared by other national leaders testifying before the House Committee on Public Works. Then Secretary of Interior, Walter J. Hickel, testified that ". . . Any management system must deal with the coastal zone in its entirety",[18] emphasizing state leadership because of their wide geographic coverage, the traditional state authorities, inherent powers to zone, and that states were free from short-term economic pressures but closer to the problems than the federal government.[19] Russell Train, Under Secretary of Interior to Hickel added, ". . . The purpose of the legislation is to give the states, as I have said, a very free hand, both in the definition of the coastal zone and in the development of the plan itself."[20] While the legislation was plagued with discussion and maneuvering on where in the federal government the new program would be housed, several concepts received wide ranging support.

1. States were to be the key element in implementing a national coastal management program. Their participation was to be on a voluntary basis.
2. States would be given latitude in designing management programs meeting general guidance for federal approval.
3. Programs were to address the entire array of coastal issues from economic development to environmental protection striving for a balanced use of the coast.
4. State programs were to coordinate the diverse interests of state agencies and local government as well as those of federal agencies in designing a program.
5. When approval was achieved, states were to be given the authority to require that activities of the federal government be consistent with the state's program.

4. Land Use Policy Phase

The land use policy is intermingled in the oceans development phase and gives CZMA the context of coastal zone as "an integral part of the entire land mass."[21] In 1970 Senator Jackson of Washington offered his first national land use bill to Congress.[22] Simply stated, this bill and subsequent bills addressed the need for establishing a national policy on the use of the land and better land management.

This highly controversial legislative effort led to over 200 land use proposals before 13 separate Congressional committees by January, 1972.[23] The effects on the coastal zone management legislation were twofold according to Zile. One, the broader land use bills encompassed the more specific concerns of the coast. Two, the numerous bills under consideration confused and delayed the legislative process.[24]

The land use bills also had impacts on CZMA which proved to be beneficial to the coastal zone management effort. One, the land use bills took the heat of political and public debate. While eyes were focused on land use, the coastal zone management legislation moved slowly but steadily throughout the Congress. When the land use effort ended in grand defeat, CZMA passed the Senate unanimously 68 to 0.[25] Secondly, the environmental movement and prospects of land use legislation drew capable young people into the planning field. Sporting a land use bias, many turned to coastal management when the land use phase ended. As a result, state programs drew from competent and motivated young planners. CZMA was where the action was.

The concepts evolved through development of coastal zone management legislation were new to federal-state relationships and to the very idea of resource management. Could a national program be made up of 35 individual state and territorial programs, each potentially different? What if states chose not to participate? Could states really exercise authority over federal agencies? Could a comprehensive management program both protect and develop the coast? While many questions must have remained unanswered in the minds of many, the Coastal Zone Management Act passed the House of Representatives 261 to 112 and the Senate with an overwhelming vote of 68 to 0.[26]

President Nixon signed the Act on October 27, 1972 with some trepidation. President Nixon commented during the signing that the ". . . coastal zones are not the only areas which need this sort of long-range attention", and continued to urge passage of his more comprehensive National Land Use Policy Act. Secondly, the President faulted the location of the program in the Department of Commerce rather than the Department of Interior as he would have preferred.[27] CZMA began with a lack of support from the President and his administration, and before it ever began President Nixon impounded the funds authorized under the Act and allocated by Congress.

B. State Legislation — Concurrent Development

The same forces which were raising the public awareness and concern for coastal management at a national level were equally active in many state legislatures. Protection of the land-water interface became a growing concern during the 1960s. Among the earliest state efforts to manage their coasts were those of California (1965) and Wisconsin (1966).

1. California — 1965

In California the initial focus was the beautiful San Francisco Bay. Concerned over the continuing filling of the Bay, the state legislature established the San Francisco Bay Conservation and Development Commission (BCDC) and charged it with developing a plan to protect the Bay from unnecessary filling.[28] As its name implies, BCDC, as it is popularly known, was not solely an environmental program but sought the rational growth for the Bay area. BCDC was a multigovernment body attempting to manage the resource as a whole; concepts later included in the federal legislation and most state programs.

2. Wisconsin — 1966

Wisconsin's action in 1966 had more of an environmental protection orientation and was motivated by concerns for the 15,000 small inland lakes, as well as the two Great Lakes. Wisconsin, however, took a more decisive step and required that rural shorelands of all rivers and lakes be zoned, setting mandatory standards for setback, development, clearing, and grading. Wisconsin's coastal management program is based on this law and 32 other separate statues most of which existed prior to 1972[29] which control a variety of uses, including removal of materials from lake beds, discharge of effluent, fisheries, power plant siting, and flood prone areas.

3. Michigan — 1970

Many states, like the federal government, traditionally addressed citizen concerns on a single-issue basis. Laws were passed to resolve the current problems. The State of Michigan, with jurisdiction over waters of four Great Lakes, cites 16 separate authorities in managing its coastal zone, 14 of which preceded the federal CZMA. Early legislative acts in Michigan redressed the concern for industrial and commercial wastewater (Waterworks and Sewage Treatment System Act — 1913 and Water Resource Commission Act — 1929) and activities on the lake beds themselves (Oil and Gas Wells Act — 1939 and Submerged Lands Act — 1955). In the 1960s it was land use control and environmental quality (Flood Encroachment Act — 1968, Subdivision Control Act — 1967, Air Pollution Control Act — 1965, and Solid Waste Management Act — 1965). Resource protection was the concern of the 1970s. (Shoreland Protection and Management Act — 1970, Mine Reclamation Act — 1970, Natural Rivers Act — 1970, Environmental Protection Act — 1970, Erosion and Sediment Control Act — 1972, Sand Dune Protection and Management Act — 1976, and Endangered Species Act — 1974).[30]

Michigan's history is typical of most legislative bodies, state or federal. Laws are passed in response to specific public concerns of the day. Legislatures and legislators can more easily understand singular issues, discuss them, set standards, and assign responsibilities. To legislate comprehensive programs is more difficult. Despite the inherently difficult nature of comprehensive programs, those crossing the lines of tradiational managing agencies, several states attempted to do so.

4. Rhode Island — 1971

One of the most far-reaching state coastal management authorities was that given to the Rhode Island Coastal Resources Management Council. Established in 1971, the Council is the "principle mechanism for management of the state's coastal resources" with permitting authority over all activities from high tide seaward.[31] Additionally, its authority extends over power plants, chemical and petroleum processing facilities, mineral extraction, shoreland protection facilities, sewerage treatment, and solid waste disposal anywhere if: (1) there is a conflict with the coastal management plan; (2) the facility would cause significant damage to the coast; or (3) the facility would make coastal areas unsuited for the use allocated.[32] The Rhode Island Program operates under the authorities of this one Act.

5. Washington — 1971

The State of Washington, also in 1971, passed the Shoreland Management Act. "The legislature finds that the shorelines of the state are among the most valuable and fragile of its natural resources and that there is great concern throughout the state relating to their utilization, protection, restoration and preservation.[33] There is, therefore, a clear and urgent demand for a planned, rational, and concerted effort, jointly performed by federal state and local governments, to prevent the inherent harm in an uncoordinated and piecemeal development of the state's shorelines."[33]

Under the Act, the State Department of Ecology was to identify shoreland of statewide significance which the legislature had defined in exhaustive detail. Local governments were to: (1) conduct a comprehensive inventory of the shoreline; and (2) regulate shoreline use under state review and approval for consistency with local plans.[34]

6. Oregon — 1971

The Oregon Coastal Conservation and Development Commission (OCCDC) was created in 1971 for the purpose of preparing a management program for coastal resources conservation and development. During its four years of existence, OCCDC developed coastal land and water policies, inventoried coastal resources and developed methods for implementing a comprehensive program. Actual implementation is carried out under the authorities of the broader Land Conservation and Development Commission. This 1973 law establishes state policy for resources management and requires local governments to develop plans and ordinances in conformance to state goals and guidelines.[35]

The Oregon program is built upon the basic authorities of the 1973 Land Use Law but additionally relies on the authorities of more than ten other agencies.[36]

7. California — 1972

In 1972 California passed one of its citizen initiatives, Proposition 20, and established the California Coastal Conservation Act. Under the Act, state and regional commissions were to plan for and regulate development along the 1,072 mi of California coast. This public mandate provided only a temporary four-year program.[37] The California plan was completed subsequently in the fall of 1975 and presented to the Governor and Legislature. Legislative action on January 1, 1977 made the Act permanent giving land use decisions to local governments with appeal to the Regional and State Commissions. The California Coastal Act of 1976 is the foundation of California's program complemented by a Coastal Conservancy Act (1976), an Urban and Coastal Park Board Act (1976), and the Coastal Commissions regulations.[38]

The states cited here are not inclusive of all actions at the state level to better manage the coastal zone. They are representatives of state initiatives prior to enactment of the federal CZMA. In the six examples given above, all had laws and programs instituted in the state prior to the enactment of CZMA and each is different. It is important in attempting to understand the national coastal management effort that with enactment of the federal law in October of 1972 many states already had tools in place to better manage the coastal zone. The federal program jumped into the middle, providing an incentive to those states who had not yet initiated management efforts and a promise of additional support and control over federal actions to those who have already taken the initiative. In fact, of the 25 state programs approved by October of 1980, 12 are based on authorities which preceded the CZMA.[39]

III. ASSESSMENTS AND AMENDMENTS

During its short nine-year existence the Coastal Zone Management Act has been amended four times, verbally discussed, formally evaluated and judicially scrutinized by untold numbers. Its struggle through Congress to enactment only began the continuing debate between Congress and the Executive. Interest groups each perceived the Act differently and continue to debate whether or not the Act is effective.

A. 1974 — Adjustment of Traditional Relationships

". . . the Coastal Zone Management Program has significant implications for federal activities as well as for state and local efforts calling for the adjustments of traditional relationships and for a new level of more effective cooperation and coordination between various levels of government."[40]

This observation was offered by Robert W. Knecht, Director, Office of Coastal Zone Management, NOAA in reporting on an NOAA-sponsored conference in March, 1974. The conference, entitled "The Coastal Imperative: Developing a National Perspective for Coastal Decision Making", drew 140 state legislators, 45 representatives from 11 environmental groups, 34 private industry representatives, 50 federal officials, 20 university researchers, 30 state officials, and numerous others. In total, over 450 people[41] attended, all trying to understand this Act that would finally be implemented.

To Robert M. White, Administrator of NOAAA, CZMA was ". . . encouraging to states to accept responsibilities to develop and implement rational management programs".[42] To the Department of Interior, it was the protection of the estuaries and natural resources.[43] The Federal Energy Office came to the conference to ". . . stress the national interest in national resource development including an energy dimension".[44] The Corps of Engineers' view was to keep the nation's ports and waterways operating.[45] Each federal agency, as one would expect, saw their interest as a major factor in each states' program.

The private sector saw CZMA as bringing ". . . more certainty to existing regulatory schemes in the States. . ."[46] and "guaranteed sites for future use and expansion."[47]

Not so, said the environmentalist who considered that CZMA ". . . was enacted primarily for environmental and protective reasons rather than for development reasons."[48]

Yes but, "The prime priority of Coastal Zone Management must be the realization of that (energy sufficiency) potential through offshore development, stabilization of policy, sitting of refineries, and permitting America to participate in the world's new deep water terminal technology."[49]

Congress, from the perspective of the National Resource Defense Council ". . . saw better management as a particular means to preserve and protect the biological and physical resources of the coastal zone."[50]

And so went the debate. Each particpant had a different view of CZMA and what it was to do. Perhaps the most accurate comment was that of Elvis J. Stahr, President, National Audubon Society who described CZMA as ". . . the schizophrenic piece of legislation that brings us together here . . . the Coastal Zone Management Act of 1972."[51]

The "Act of 1972" was already seeing its first amendment through the legislative process. The amendment, passed enthusiastically, provided technical changes to the basic Act by increasing authorized funding levels and extending the time allowed to develop programs. Congress had not lost its eagerness for CZMA. "The fact that 31 of 34 states have joined the program in its first year demonstrates their enthusiastic belief in its value."[52]

B. 1976 — An Uncertain Future

By December of 1976 the Comptroller General of the U.S. had labeled the Coastal Management Program with the tag "An Uncertain Future".[53] Barely two years after it began, the General Accounting Office conducted a lengthy review of the Coastal Management Program. (Funding was delayed until 1974 because funds were impounded by President Nixon.) GAO observed delay in program development, a need for increased federal monitoring of state programs, limited participation by federal agencies, and an uncertainty of the federal consistency provisions as major problem areas for Coastal Management. While GAO recognized the unique approach to federal-state-local relationships and the change of political climate between 1972 and 1976, their report was less than supportive.

States felt that they were making good progress, according to the GAO survey of 30 states. GAO, however, felt programs were slow to develop because states: (1) lacked experience; (2) were depending upon local plan development; (3) were trying to tie together existing laws; (4) were meeting strong public resistance; and (5) lacked political support.[54] Federal agency participation in state program development was limited. According to the GAO review, federal agencies had only recently recognized their responsibilities under CZMA and lacked mechanisms to participate in state program development.[55]

All was not well with CZMA. Interest groups which had viewed CZMA as the answer to their specific problems were becoming disenchanted as the states and the federal office of Coastal Zone Management sought programs that managed not one but a variety of coastal concerns. The time needed to develop programs was more than Congress had anticipated and, federal agency participation in state program development was limited.[56]

Recognizing the problems encountered and the need for improved clarity, Congress amended the Act again in July of 1976. Funding authority was extended from 1977 to 1980 with an increased federal share of program cost, from 66 and 1/2% to 80%. Three new requirements were added for program approval: planning processes of shoreline erosion control, energy facility planning, and beach access. Two procedures were improved: (1) a provision of an interim grant between program development and final approval; and (2) a mediation process to resolve federal/state disagreements. Added to CZMA was a $1.2 billion Coastal Energy Impact Fund (Section 308) to aid states in accommodating new energy facilities.[57] (See Figure 1).

C. 1978 — Losing its Constituency

By 1978 those who had rallied around the Coastal banner in 1972 were beginning to drift away. Coastal Management was losing its constituency or so it appeared to many. "I would have to say that I do not believe that a constituency for Coastal Zone Management exists in this country today to the degree necessary to have this program continue effectively in the future . . .," said John Hussey, Chairman of the Department of Commerce's Coastal Zone Management Advisory Committee.[58] William Reilly, President of the Conservation Foundation, summed up the attitude in early 1978, "We now have discovered that the environmental road is much longer and more difficult than we had thought . . ."[59] Ecologists, reported Reilly, believe that the program has fallen into the hands of the land use planners. The planner regrets that the state is turning back to local governments for program implementation. Local governments feel that the states are taking the major share of the funds and intruding upon local autonomy. The developer sees more government and a failure with programs to designate specific sites for future development.[60]

Given the charge by Congress to develop comprehensive programs, Coastal Management

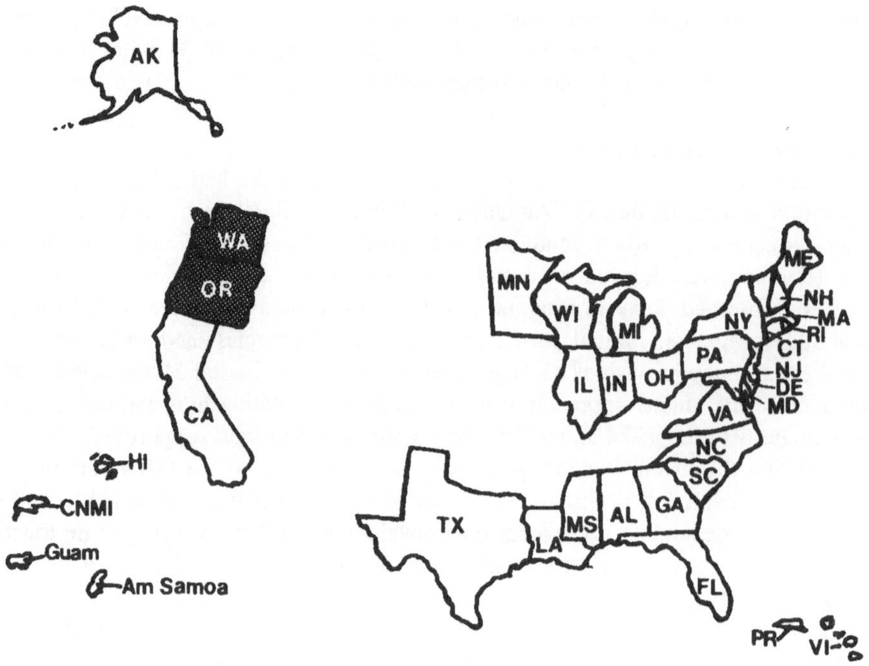

FIGURE 1. Programs approved in 1976 and 1977.

was destined to lose at least some of its initial supporters. CZMA was not a wetland protection act which would preclude all further development from wetlands. It was not a facility siting act which would set aside shoreland for future development. It was not a fisheries management program, nor a local assistance program, nor a harbor improvement program. It was intended to respond to all of these concerns and more. As a result, none of the special interest area advocates found that CZMA was totally fulfilling their needs. It wasn't intended to. But was it achieving its purpose?

According to Bob Knecht, the Acting Assistant Administrator for Coastal Zone Management, NOAA, "Yes, but . . ."[61] Knecht, under whose able leadership the federal program was launched, recognized the political reality of developing state programs. In March of 1978, 30 states and 4 of the 5 territories were participating in the federal program and had begun the process of developing an approvable program.[62] Because of the federal government's late entry into Coastal Management, many states had programs which would meet the federal standards. Others were progressing satisfactorily. While state programs were moving at an acceptable rate, "In general they will be in need of additional refinements and elaborations to make the program fully effective in a comprehensive coastal management sense", said Knecht.[63]

In late 1978, the Coastal Zone Management Advisory Committee, describing CZMA as "an experiment in intergovernmental relations", recommended a gradual phase out of federal dollars.[64] Critical of the implementation of CZMA, the Advisory Committee, appointed by the Secretary of Commerce, found a lack of understanding and serious problems with federal consistency as major concerns. The Committee further called for a public initiative as a first major step to improving program implementation and recommended more definite performance standards.[65]

By the close of the year, 13 state programs had been approved by the Secretary of Commerce.[66] The experiment, however, was with state government programs. "State Coastal

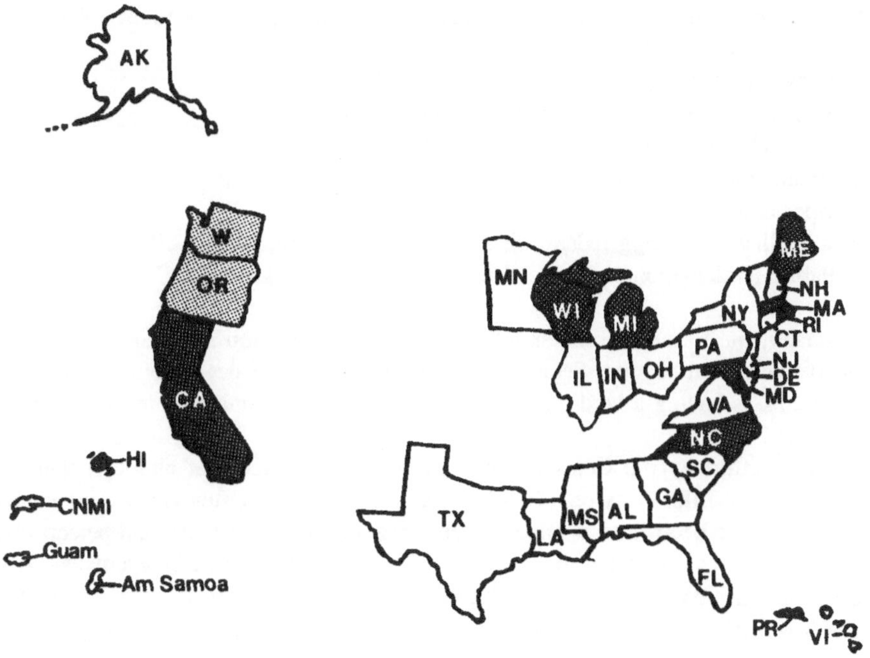

FIGURE 2. Programs approved in 1978.

Management are supposed to pull together not only the multiple concerns of the state but also those of numerous federal agencies, the development ones and the protective ones, reconciling virtually every possible demand on the Coastal Zone from transportation, to recreation, to wildlife.''[67] Meanwhile, back in Washington, D.C., 37 subagencies, 9 independent agencies, and 5 departments all had a say in national oceans policy.[68] No one agency had the responsibility or the authority to express national oceans policy or to coordinate federal ocean activities. Even in utopia, 35 state and territorial programs could not write national policy and, without policy guidance, each federal agency continued to write its own. The need for national policy on oceans and coastal resource management was beginning to become increasingly apparent (Figure 2).

D. 1979 — Encouraging Signs

The states, concerned that Congress understand the prevailing attitude on CZMA, called for a national forum in 1979 to develop a consensus position on coastal management. The Coastal States Organization, an affiliate of the National Governors' Conference, called on the leaders of industry, public interest groups, and government to participate in the discussion ''Options for the 80's''. Representatives of the petroleum industry, environmental groups, harbor commissions, governmental officials, and others debated whether CZMA was fulfilling its promise.[69] After two days of discussion, the 75 participants representing interest as diverse as the National Resources Defense Council and Exxon Company, U.S.A., reached a consensus on several major points.

1. Despite this great variety in expectations, there is a strong consensus that the management program has made a positive contribution to the rational management of coastal resources and should continue to receive support.
2. The forum recognized the difficulties in initiating coastal management programs and recognized the early stages of most activities recommending that more time was needed before judgments of success or failure were made.

3. Finally, the forum recognized that changes were needed in the future. Six subjects were recommended for specific action:
   a. National interest ". . . clarification is needed . . ."
   b. Federal consistency ". . . not a strong incentive . . ."
   ". . . clarification and simplification is needed."
   c. Evaluation — strengthen Section 312.
   d. Special areas management ". . . time to begin use . . ."
   e. Coordination among federal agencies ". . . must improve . . ."
   f. Public participation and education ". . . the need is great . . ."

Additionally, the forum recommended that funding for Section 306 should continue for the next five years and beyond. Section 305 funds should terminate. As others, they chided CZM programs as having a land-planner bias with insufficient attention to the "wet" side of the Coastal Zone.[70]

This unique effort by the Coastal State Organization showed that although there were some fundamental differences on CZMA there were also many fundamental agreements. CZMA continued to be viewed as an important program from many different persepectives.

As the assessments continued the federal Office of Coastal Zone Management (OCZM) offered its own entitled "The First Five Years of Coastal Management". OCZM concluded that CZMA was both unique and ambitious.[71] Assessing its results quantitatively, the report cites new institutions created and miles of shoreline now encompassed in approved state programs. By January of 1979, 13 states had received approval covering 1/3 of the nation's coast and almost 50% of the coastal population.[72] "Encouraging signs . . ." notes the report, "too early, but beginning to make a difference." The report focuses only on the state program efforts. ". . . the effectiveness of the overall national effort rests almost entirely on the quality and accomplishments of each state's coastal management efforts", said Robert W. Knecht, Acting Administrator for Coastal Zone Management, NOAA.[73] The national coastal management effort continued to be perceived by the federal office as the sum of 35 state programs.

President Carter offered a glimmer of hope for the coordination of federal programs and the establishment of national coastal policies. In his environmental message, August 2, 1979, the President directed the Secretary of Commerce ". . . to conduct a systematic review of federal programs that significantly effect coastal resources, . . . (to) provide the basis for specific recommendations to improve federal actions affecting the coastal zone and to develop any additional legislation needed to achieve our national coastal management goals"[74] (Figure 3).

E. 1980 — Year of the Coasts

Spurred by a coalition of interest groups, the Coastal Alliance, the President proclaimed 1980 as the "Year of the Coast". Numerous governors issued paralleling proclamations and many interested groups joined the effort. A national focus was indeed given to oceanic and Great Lakes shorelines. Sand castle building contests, coastal walks, educational programs, news media coverage of coastal issues, coastal T-shirts, bumper stickers, and pins all aided in increasing the public's awareness and understanding of the coast and it was fun! After six years of struggling with an ambiguous Act and differing perspectives on what CZMA was, the Year of the Coast reminded all of the common concern; the coast. The business at hand in 1980 was reauthorization, and as Congress was considering the form it should take, the General Accounting Office offered its second evaluation to the Secretary of Commerce.

This time GAO entitled their report "Problems Continue in the Federal Management of the Coastal Management Program". "GAO found during this review that many of the same

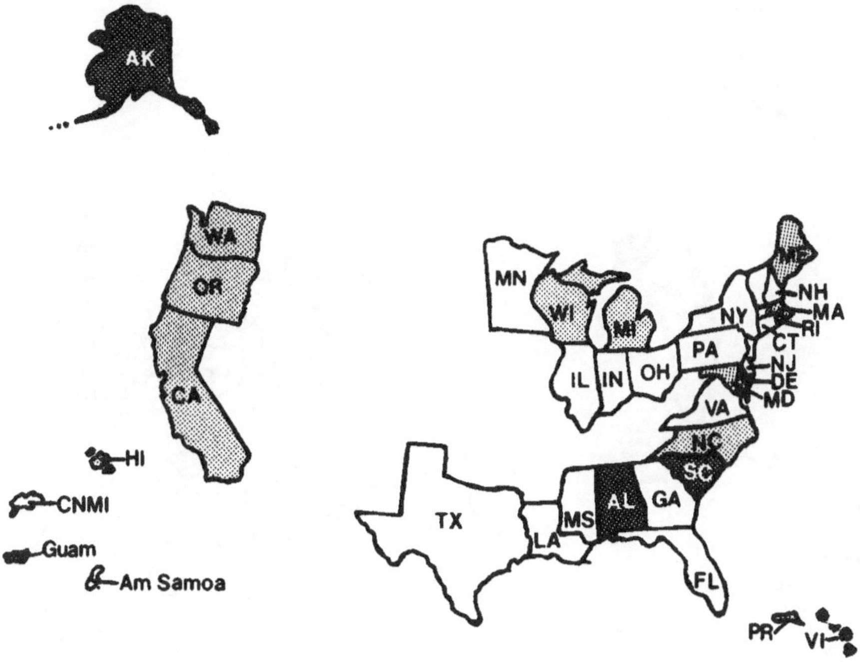

FIGURE 3. Programs approved in 1979.

problems cited in the previous report continue to exist.''[75] GAO again called for the establishment of evaluation guidelines and criteria to provide a more systematic approach in evaluating state programs.[76] The emphasis of the recommendations was a stronger federal role in evaluating and monitoring state programs. Labeling Congressional intent as ''NOAA's management philosophy'', the report identifies the latitude provided by NOAA to states in the design and implementation of programs and points out Secretary of Commerce's role to promulgate rules, coordinate programs with federal agencies, and continually review states' performance.[77] GAO's attitude and opinions of CZMA had changed little in four years. From the report, it does not appear that GAO agrees with the strong state role that Congress had initiated in 1972 and prefers instead a more dominate federal role.

Perhaps one of the most straightforward assessments of CZMA in 1980 was provided by the Conservation Foundation. Their report suggests to environmentalists that ''. . . they should recognize that more narrowly focused programs may do a better job of protecting resources'',[78] and recommended that efforts be directed at state and local governments and less to Washington, D.C. After six years of CZMA, interest groups were beginning to appreciate that the state programs would not, and could not, focus on any one area of interest. ''The program, moreover, is a management program intended to achieve multiple and even conflicting purposes.''[79] This understanding reflects maturing relationships between CZMA and its constituents. Perhaps the ''lack of support'' suggested by the National Advisory Committee on Coastal Zone Management in 1976 was returning. Or perhaps it never left, but only needed time to understand CZMA and observe its results. By May of 1980, 17 of 19 approved programs were less than two years old and little experience had been gained in the practice of coastal management. ''It's only within the last year that we have been able to pay attention to the coast instead of federal regulations.''[80]

Under the diligent and able leadership of Representative Studds of Massachusetts (Chairman, Subcommittee on Oceanography), Congress undertook its own evaluation. Leaving Washington, D.C. to receive testimony at six different locations across the country (New

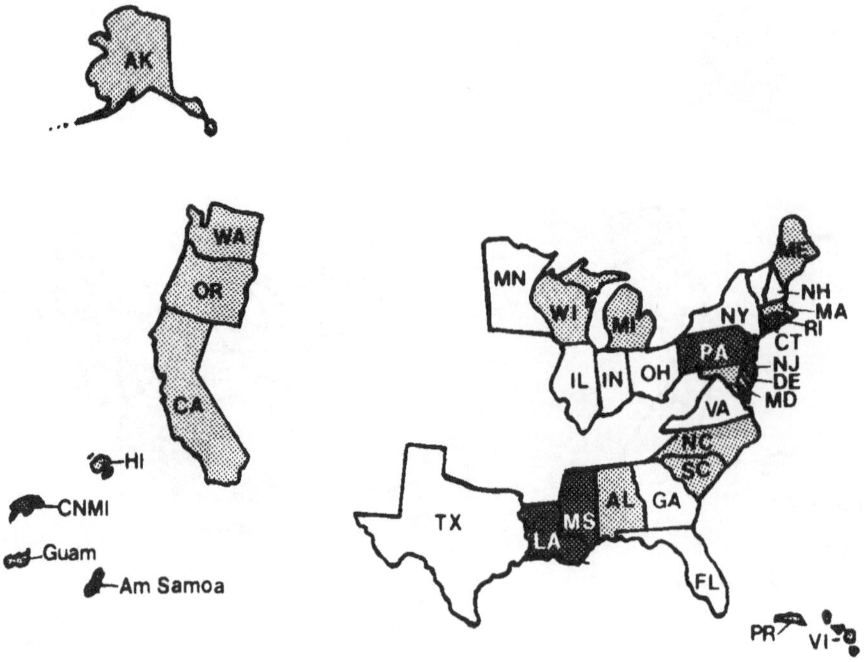

FIGURE 4. Programs approved in 1980.

Orleans, Detroit, Seattle, San Francisco, Atlantic City, and Boston), the Subcommittee received extensive support for CZMA. Subcommittee member Bonior of Michigan reported to the House ". . . I have rarely seen a piece of legislation greeted with so much enthusiasm as I have this one."[81]

In the closing hours of the 96th Congress, the Coastal Zone Management Improvement Act of 1980 passed both Houses of Congress by unanimous consent. The Act provides significant changes to CZMA.

1. It extends authorization until 1985.
2. It provides additional language to the Congressional Declaration of Policy in an effort to define the national interest.
3. It adds a new grant section to promote low-cost construction and small scale development.
4. It adds coal, and alternative ocean energy activities, to the Coastal Energy Impact Program.
5. It authorizes grants for interstate efforts.
6. It strengthens the review of performance provisions.
7. It adds island preservation and deletes beach access from the Estuarine Sanctuary section.
8. It terminates the Coastal Zone Management Advisory Committee.
9. It requires a biennial report to the President.
10. It provides 60 days for Congressional review of final regulations.[82]

The Year of the Coast ended with new life for CZMA. Twenty-five states had achieved federal approval covering 75,000 mi (78%) of the nation's coast. Six states' programs were in final stages of development and four states were no longer participating in the federal programs (Figure 4).

F. 1981 — Achieved Its Purpose

"Funding for the State Coastal Zone Management and Sea Grant Colleges Grant Programs would be terminated because they have largely achieved their original purpose to develop state and local capabilities for addressing marine and coastal issues."[83]

While the Congressional debate on reauthorization had proceeded through months of discussion and 11 pages of the Congressional record, the Executive Department's third evaluation took just one sentence in the Fiscal Year 1982 Budget Revisions.

The rationale for termination of CZMA is interesting. Of the 25 states approved, 12 developed state capabilities prior to the enactment of CZMA. Congressional intent and CZMA of 1972 authorizes both development grants (Section 305) and administrative grants (Section 306). Congress has historically placed its emphasis on implementation (Section 306) by originally authorizing funding levels of $9 million for development and $30 million for implementation.[84] The 1980 amendments authorize no funds for development $48 million for implementation and $20 million for resource management grants.[85] Had Congress intended only to develop capabilties and not implement programs, there would have been little incentive for these states to participate in the national program. With the administration's proposed budget reductions, one incentive for state participation was being seriously threatened.

The other incentive, consistency of federal agency actions with state programs, was also being threatened. In July 1980, The California Coastal Commission requested the Secretary of the Department of Interior to submit a consistency determination on the proposed Outer Continental Shelf Lease Sale No. 53. (Lease Sale No. 53 contained 243 tracts, 150 of which were in the Santa Maria Basin.) In October, the Department of Interior notified the Commission, "that the preleasing activities had no 'direct effects' on California's coastal zone." California responded that the deletion of 29 tracts in the Santa Maria Basin was necessary in order to be consistent with the California coastal management plan. In April 1981, the Secretary of Interior, James Watt, announced his intent to lease the entire Santa Maria Basin because of the "overriding national interest".

California sued Secretary Watt in April.[86] Paralleling the California suit against Watt was a similar suit entered into by the Natural Resources Defense Council, The Sierra Club, Friends of the Earth, Friends of the Sea Otter, and the Environmental Coalition. As party to the state's suit were 16 California counties and cities. Eleven petroleum companies intervened in behalf of the Department of Interior.[87] Additionally, the Coastal States Organization and the states of Alaska, Oregon, South Carolina, Maine, and Hawaii all joined California in amicus. California was granted a preliminary injunction in May, and on July 10, 1981, the United States District Court, Central District of California, found in favor of the California group claim that Interior's action was inconsistent under the Coastal Zone Management Act.[88]

Back in Washington, D.C., Secretary Watt wrote to Secretary Malcolm Baldridge, Secretary of Commerce, expressing his objection to NOAA's preliminary regulations which indicated prelease activities as "directly affecting" the Coastal Zone. Soon after, the Department of Commerce decided to finalize its regulations regarding consistency.[89] In an abrupt about-face, Commerce identified prelease activities as not directly affecting the Coast and therefore, exempt from a consistency review. With the issuance of final regulations in July, California again filed suit, this time against the Department of Commerce.[90]

In July, North Carolina initiated action against Secretary Watt over Lease Sale No. 56, charging that Interior had failed to file a determination that the sale is consistent with the state's coastal policies.[91]

Whether federal consistency will survive through the court and/or changes in regulation is yet to be seen. The second federal incentive, however, is also clearly in jeopardy.

Congress, meanwhile, worked on restoring the financial incentive. In oversight hearings before the House Subcommittee on Oceanography on March 31, 1982, 19 people testified supporting the return of federal funding, at least to provide a reasonable program phasedown in lieu of the abrupt termination of the program proposed in the President's budget. Coastal States Organization Chairman, Jim Ross of Oregon, advised the Committee that, "Only four states have indicated *any* possibility of being able to cover the loss of federal funding at this time. Thirteen states said their CZMA programs would be 'shut down' if CZM funds were completely withdrawn during Fiscal year 1982."[92] Convinced of the need for at least a systematic phasedown, $40 million of rescinded CEIP funds were reprogrammed by Congress for the phasedown.[93]

In response to President Carter's environmental message of October 1979, the Department of Commerce submitted its report on the review of federal programs to the President in January 1981. A total of 33 recommendations were provided suggesting changes in the way federal government conducts its business on the coast.[94] The report is limited to five broad areas:

1. Programs providing development and reconstruction assistance in coastal hazard areas
2. Programs supporting infra-structure development
3. Access to federal lands
4. Programs related to energy development
5. Improved coordination for planning and permitting in special coastal areas[95]

Finally, an attempt was being made to coordinate federal coastal activities. However, a new President was in office and the administration had changed. In transmitting the report to interested parties, Knecht noted that to date the report ". . . has not been reviewed or endorsed by the current administration; . . ." but that review was expected in the coming months.[96]

In August of 1981, the CZMA's future was clouded. Long-term funding (federal or state) was questionable, consistency undetermined, and the coordination of federal activities unpursued.

IV. AN UNFINISHED AGENDA

A. National Program — Never Achieved

A National Coastal Management Program has never been developed, and only with strong leadership of both the executive and legislative branches of government can one be achieved.

In 1972, states were seen as "the key" element in implementing a National Coastal Management Program. In hindsight, perhaps the words should have been "a key". Twenty-five states have approved coastal programs covering 75,000 mi of shoreline (78% of the total coast) and 60% of the coastal population. A significant improvement over 1972, but still missing a number of important states. The South Atlantic is without the states of Georgia and Florida, the Gulf without the state of Texas, and the Great Lakes without Minnesota, Illinois, Indiana, Ohio, and New York.

In 1978, Congressman Breaux told participants in "Coastal Zone - 78" that the U.S. Government has 37 subagencies, 9 independent agencies, and 5 departments, each having a say in ocean policy. "Thus, we have reached a point in oceans management where we are so overlapping in our respective areas of expertise, or areas of jurisdiction that we are all competing."[97]

Three years later, James W. Curlin, Deputy Assistant Secretary for Land and Water Resources, Department of Interior, told the Marine Technology Society that the new Reagan Administration would find:

- A fragmented assortment of ocean-related programs scattered among 10 departments,

8 independent agencies, and 38 agencies or subagencies, and lacking any effective means to coordinate or guide government actions

- An absence of any identifiable objectives or guidelines to focus ocean policy or to measure progress
- A phalanx of ocean-related legislation, some of it outdated, some of it ill-conceived, and much of it poorly implemented by the agencies
- An ocean policy whose major components are strongly oriented toward environmental protection without a similar agenda to provide for balanced economic development of marine resources
- A depreciated capacity to perform quality ocean science, research, and development within the government
- A LOS treaty which, although awaiting final initialling, is still seriously flawed and unlikely to pass Senate muster
- A Coast Guard that is understaffed, underbudgeted, and too overcommitted to effectively carry out the missions assigned to it by Congress[98]

Others were also publicly advising the new President. "The President must see that the dozen or more different segments of the federal bureaucracy which now handle ocean matters must get clear direction from the top (from the White House, that is). Otherwise, each of those agencies, departments or bureaus will go its own way and produce, even as we have today, duplicative and contradictory policies. Policies which are self-defeating in the sense that nothing of much significance ever gets done," said John R. Botzum, Editor of the Nautilus Press in a speech entitled "A National Ocean Policy — Now!".[99]

The need for clear direction to federal agencies in the forms of policies on coastal and oceans management has not yet been achieved. Nor has total state participation in the federal Coastal Management Program been achieved. Neither federal, state nor local government policies or improved management practices can individually make up a national program, even with 100% participation. What is needed for a national program is the participation of all three levels of government and clear policy direction. What we have today in Coastal Management is not a national program but 25 state programs meeting federal standards and 40 or more federal agencies each doing its own thing.

B. Federal Programs — Going To Sea

The federal government's management focus is moving away from the shoreline and out to sea and will strongly emphasize resource development.

The Reagan Administration quickly made its policies on coastal zone management known, by terminating funding for the program. The focus of the Reagan Administration is economic recovery and the oceans policies of the administration are strongly oriented to that goal.

Secretary of Commerce Malcolm Baldridge provided the administration's oceans policies in April of 1981. Noting a $50 billion contribution to the Gross National Product (GNP) from ocean-related activities, he then delineated the specific objectives of the Reagan Administration.

- Marine Fisheries Development ". . . stimulate the economic viability of the fishing industry and reduce the reliance on foreign fish products."
- Offshore Oil and Gas Extraction ". . . accelerate the leasing of federal lands on the outer continental shelf to reduce reliance on foreing import and strengthen the U.S. economy."
- Marine Mineral Exploitation ". . . develop a legal and regulatory framework that encourages private investments in seabed mining."
- The Ocean and National Energy Policy - "NOAA is now creating a legal and regulatory

framework to guide development of OTEC (Ocean Thermal Energy Conservation) system by private industry. "The Department of Commerce and other federal agencies will be examining ways to promote expansion of coal transportation by ships."

- State and Local Roles in Coastal Programs ". . . Programs of research and resource management planning should be maintained at the state and local level. "As we look to the 1980's, I expect that state and local government will continue the work stimulated by these federal programs and direct federal involvement can be reduced."[100]

Clearly the Administration's view and its efforts will be oriented toward resource development, development which will heavily impact the shoreland. One needs to keep in mind, however, that this viewpoint may or may not be shared by Congress. Over its history, Congress and the President have not often agreed on the subject of coastal management.

C. State Programs — Hopeful

State coastal management programs will continue; some, however, will be terminated and most will operate at greatly-reduced levels due to the existing economic conditions.

The need for improved management of the coast has not diminished; if anything the increasing population and development pressures have intensified the need. Current state programs are funded at an average level of 77% federal funds, 23% state or local funds. Of the federal funds received, 40% goes to local units of government. When surveyed by the Coastal States Organization in March of 1981, 22 of 26 states responded that the political/practical capability of their state covering the loss of federal funding was poor (8) or extremely poor (14). Phasedown of federal funding, however, would alter program continuation. Eighteen of 25 states reported the phasedown would make a "significant difference" with 5 more noting it would make "some differerence".[101]

Of the 25 states which have achieved program approval, only 2 have been operating for more than 4 years. Ten states now have over three year's experience, six over two year's experience, and seven with less than one year.

The actual changes in management, e.g., permit consolidation, and changes in shoreland use are just beginning to be seen in the older programs. More recently approved programs have yet to fully demonstrate their usefulness to the state and local governments.

The uncertainty of federal incentives, both financial assistance and federal consistency at this critical point in time, combined with the economic problems of state and local governments, will most likely have a devastating impact on the management of the coast of this nation.

V. TOMORROW

We have not yet completed the job of improving the comprehensive management systems which the Stratton Commission and Congress told us was so vital to the health of this nation's coasts. Less than 75% of the states have adopted and implemented comprehensive coastal management programs with several key states having dropped out of the program. We have no federal coastal zone management policy. Instead we have 40 or more federal agency's policies, many of which are incompatible. As a result, while we have made improvements since 1972, we still do not have a national coastal zone management program.

As a nation we will look more and more to the water. The need for food, oil and gas, minerals, or the improved transportation of commodities to foreign markets will make the oceans and the Great Lakes increasingly important. Such demands, now spurred on by the federal government in turn place increased pressures on the land-water interface. Secretary Baldridge has identified increased development of the ocean resources as an objective of this administration. It is illogical then for the federal government to disassociate itself with any responsibility for managing the resulting impacts.

The economic hardships of the nation are being felt everywhere and by all three levels of government. Shifting the fiscal cost for management of the coast from one level to another does not solve the problem nor does it remove the fiscal burden from the source of government revenues, the taxpayer. Two things have to happen. (1) a set of national policies must be developed; and (2) a stable source of funding must be established and available to all levels of government.

First, truly national, and durable, policies can only be established by the Congress. Renewed attention should be given to the national oceans policies committee to establish this national policy framework. Input should be sought from all levels of government, from the private sector, and from interest groups in developing the future national direction. Federal agency policies and activities should then be made consistent with these national policies. State and local governments could then operate under a fixed set of national guidelines rather than disparate agency programs.

Secondly, the cost for managing the coast should be shared among all levels of government, federal, state, and local, and stabilized over the long-term. Each level of government has its own responsibilities in the overall management of the coast. No greater wisdom exists at any one level, only a different perspective, different priorities, different responsibilities and access to different resources. For a change to occur all three levels of government must participate in improving coastal management. The absence of a stable source of funding will cause any of the participants to only look at short-term and "completable projects" rather than long-term improvement of the management process. Sharing outer continental shelf revenues with coastal states offers the most reasonable approach to a stabilized source.

Until national policies are set and a stable funding source established, coastal management will be just another categorical grant program to be used for short-term objectives. The federal administrating agency will never obtain sufficient strength to influence any other agency's activities and agency policies will continue to be incompatible. Over the past ten years we have made considerable investment to better manage the coast and have made good progress. The job is not yet done, it has only just started.

REFERENCES

1. Legislative History of the Coastal Zone Managment Act of 1972, as Amended in 1974 and 1976 with Section-By-Section Index, Washington, U.S. Government Printing Office, Washington, D.C., 1976, 444.
2. U.S. Department of Commerce, Office of Coastal Zone Management, CZM Information Exchange, July 1981, 6.
3. Executive Office of the President, Office of Management and Budget, Fiscal Year 1982 Budget Revisions, Washington, D.C., March 1981, 49.
4. **Zile, Z. L.,** A legislative-political history of the Coastal Management Act of 1972, *Coastal Zone Manage. J.*, 1(3), 236, 1974.
5. **Zile, Z. L.,** A legislative-political history of the Coastal Management Act of 1972, *Coastal Zone: Manage. J.*, 1(3), 238, 1974.
6. **Zile, Z. L.,** A legislative-political history of the Coastal Management Act of 1972, *Coastal Zone Manage. J.*, 1 (3), 246, 1974.
7. **Zile, Z. L.,** A legislative-political history of the Coastal Management Act of 1972, *Coastal Zone Manage. J.*, 1 (3), 244, 1974.
8. **Zile, Z. L.,** A legislative-political history of the Coastal Management Act of 1972, *Coastal Zone Manage. J.*, 1 (3) 245, 1974.
9. **Zile, Z. L.,** A legislative-political history of the Coastal Management Act of 1972, *Coastal Zone Manage. J.*, 1 (3), 253, 1974.
10. **Zile, Z. L.,** A legislative-political history of the Coastal Management Act of 1972, *Coastal Zone Manage. J.*, 1 (3), 253, 1974.

11. **Zile, Z. L.**, A legislative-political history of the Coastal Management Act of 1972, *Coastal Zone Manage. J.*, 1(3), 254, 1974.
12. **Zile, Z. L.**, A legislative-political history of the Coastal Management Act of 1972, *Coastal Zone Manag. J.* 1(3) 253, 1974.
13. **Zile, Z. L.**, A legislative-political history of the Coastal Management Act of 1972, *Coastal Zone Manage. J.*, 1 (3), 253, 1974.
14. Legislative History of the Coastal Zone Management Act of 1972, as Amended in 1974 and 1976 with Section-By-Section Index, Washington, Government Printing Office, Washington, D.C., 1976, 10.
15. Legislative History of the Coastal Zone Management Act of 1972, as Amended in 1974 and 1976 with Section-by-Section Index, U.S. Government Printing Office, Washington, D.C., 1976, 1.
16. Legislative History of the Coastal Zone Management Act of 1972, as Amended in 1974 and 1976 with Section-by-Section Index, U.S. Government Printing Office, Washington, D.C., 1976, 1.
17. Legislative History of the Coastal Zone Management Act of 1972, as Amended in 1974 and 1976 with Section-by-Section Index, U.S. Government Printing Office, Washington, D.C., 1976, 15.
18. Legislative History of the Coastal Zone Management Act of 1972, as Amended in 1974 and 1976 with Section-by-Section Index, U.S. Government Printing Office, Washington, D.C., 1976, 24.
19. Legislative History of the Coastal Zone Management Act of 1972, as Amended in 1974 and 1976 with Section-by-Section Index, U.S. Government Printing Office, Washington, D.C., 1976, 26.
20. Legislative History of the Coastal Zone Management Act of 1972, as Amended in 1974 and 1976 with Section-by-Section Index, U.S. Government Printing Office, Washington, D.C., 1976, 44.
21. **Zile, Z. L.**, A Legislative-Political History of the Coastal Management Act of 1972, *Coastal Zone Manage. J.*, 1(3), 1974, 268.
22. **Zile, Z. L.**, A legislative-political history of the Coastal Management Act of 1972, *Coastal Zone Manage. J.*, 1 (3), 268, 1974.
23. **Zile, Z. L.**, A legislative-political history of the Coastal Management Act of 1972, *Coastal Zone Manag. J.*, 1 (3), 269, 1974.
24. **Zile, Z. L.**, A legislative-political history of the Coastal Management Act of 1972, *Coastal Zone Manage. J.*, 1 (3), 269, 1974.
25. Legislative History of the Coastal Zone Management Act of 1972, as Amended in 1974 and 1976 with Section-By-Section Index, Government Printing Office, Washington, D.C., 1976, 7.
26. Legislative History of the Coastal Zone Management Act of 1972, as Amended in 1974 and 1976 with Section-by-Section Index, U.S. Government Printing Office, Washington, D.C., 1976, 7.
27. Legislative History of the Coastal Zone Management Act of 1972, as Amended in 1974 and 1976 with Section-by-Section Index, U.S. Government Printing Office, Washington, D.C., 1976, 459.
28. State of California Coastal Management Program and Final Environmental Impact Statement, Department of the Interior, California, August 1977, 19.
29. State of Wisconsin Coastal Management Program and Final Environmental Impact Statement, Department of Interior, Wisconsin, August, 1977, 173.
30. State of Michigan Coastal Management Program and Final Environmental Impact Statement, Department of Interior, Michigan, March 1978, 105.
31. State of Rhode Island Costal Management Program and Final Environmental Impact Statement, Department of Interior, Rhode Island, March 1978, 1.
32. State of Rhode Island Coastal Management Program and Final Environmental Impact Statement, Department of Interior, Rhode Island, March 1978, 1.
33. Washington State Coastal Zone Management Program, U.S. Government Printing Office, Washington, D.C., June 1976, 3.
34. Washington State Coastal Zone Management Program, U.S. Government Printing Office, Washington, D.C., June 1976, 29.
35. State of Oregon Coastal Management Program, Final Environmental Impact Statement, Department of the Interior, Oregon, March 1977, 3.
36. State of Oregon Coastal Management Program, Final Environmental Impact Statement, Department of the Interior, Oregon, March 1977, 6.
37. State of California Coastal Management Program and Final Environmental Impact Statement, Department of the Interior, California, August 1977, 17.
38. State of California Coastal Management Program and Final Environmental Impact Statement, Department of the Interior, California, August 1977, 10.
39. U.S. Department of Commerce, Office of Coastal Zone Management, "Status of Federal Approval of State Coastal Zone Management Programs", October, 1980.
40. Coastal Zone Management, "The Coastal Imperative: Developing a National Perspective for Coastal Decision Making", Government Printing Office, Washington, D.C., 1974.

41. Coastal Zone Management, The Coastal Imperative: Developing a National Perspective for Coastal Decision Making, U.S. Government Printing Office, Washington, D.C., 1974, 8.
42. Coastal Zone Management, The Coastal Imperative: Developing a National Perspective for Coastal Decision Making, U.S. Government Printing Office, Washington, D.C., 1974, 17.
43. Coastal Zone Management, The Coastal Imperative: Developing a National Perspective for Coastal Decision Making, U.S. Government Printing Office, Washington, D.C., 1974, 18.
44. Coastal Zone Management, The Coastal Imperative: Developing a National Perspective for Coastal Decision Making, U.S. Government Printing Office, Washington, D.C., 1974, 27.
45. Coastal Zone Management, The Coastal Imperative: Developing a National Perspective for Coastal Decision Making, U.S. Government Printing Office, Washington, D.C., 1974, 40.
46. Coastal Zone Management, The Coastal Imperative: Developing a National Perspective for Coastal Decision Making, U.S. Government Printing Office, Washington, D.C., 1974, 65.
47. Coastal Zone Management, The Coastal Imperative: Developing a National Perspective for Coastal Decision Making, U.S. Government Printing Office, Washington, D.C., 1974, 67.
48. Coastal Zone Management, The Coastal Imperative: Developing a National Perspective for Coastal Decision Making, U.S. Government Printing Office, Washington, D.C., 1974, 70.
49. Coastal Zone Management, The Coastal Imperative: Developing a National Perspective for Coastal Decision Making, U.S. Government Printing Office, Washington, D.C., 1974, 82.
50. Coastal Zone Management, The Coastal Imperative: Developing a National Perspective for Coastal Decision Making, U.S. Government Printing Office, Washington, D.C., 1974, 99.
51. Coastal Zone Management, The Coastal Imperative: Developing a National Perspective for Coastal Decision Making, U.S. Government Printing Office, Washington, D.C., 1974, 166.
52. Legislative History of the Coastal Zone Management Act of 1972, as Amended in 1974 and 1976 with Section-By-Section Index, Government Printing office, Washington, D.C., 1976, 551.
53. Comptroller General of the United States, "The Coastal Zone Management Program: An Uncertain Future", December, 1976, 1.
54. Comptroller General of the United States, The Coastal Zone Management Program: An Uncertain Future, December, 1976, 33.
55. Comptroller General of the United States, The Coastal Zone Management Program: An Uncertain Future, December, 1976, 62.
56. Legislative History of the Coastal Zone Management Act of 1972, as Amended in 1974 and 1976 with Section-By-Section Index, Government Printing Office, Washington, D.C., 1976, 551.
57. Report of the Committee of Conference on S. 586 — Coastal Zone Management Act Amendments of 1976, Government Printing Office, Washington, D.C., June 1976.
58. Coastal Zone '78, Symp. Tech. Environ. Socioecon. and Reg. Aspects Coastal Zone Manage., San Francisco, Calif., March 14-16, 1978, Am. Society of Civil Engineers, New York, 2456.
59. Coastal Zone '78, Symp. Tech., Environ., Socioecon., and Reg. Aspects Coastal Zone Manage., San Francisco, Calif., March 14-16, 1978, Am. Society of Civil Engineers, New York, 2446.
60. Coastal Zone '78, Symp. Tech., Environ., Socioecon., and Reg. Aspects Coastal Zone Manage., San Francisco, Calif., March 14-16, 1978, Am. Society of Civil Engineers, New York, 2446.
61. Coastal Zone '78, Symp. Tech., Environ., Socioecon., and Reg. Aspects Coastal Zone Manage., San Francisco, Calif., March 14-16, 1978, Am. Society of Civil Engineers, New York, 2439.
62. Coastal Zone '78, Symp. Tech., Environ., Socioecon., and Reg. Aspects Coastal Zone Manage., San Francisco, Calif., March 14-16, 1978, Am. Society of Civil Engineers, New York, 2438.
63. Coastal Zone '78, Symp. Tech., Environ., Socioecon., and Reg. Aspects Coastal Zone Manage., San Francisco, Calif., March 14-16, 1978, Am. Society of Civil Engineers, New York, 2441.
64. Public Support for Coastal Zone Management Programs: The Implementation of the Coastal Zone Management Act of 1972, Coastal Zone Management Advisory Committee, U.S. Department of Commerce, Washington, D.C., June 1978, 25.
65. Public Support for Coastal Zone Management Programs: The Implementation of the Coastal Zone Management Act of 1972, Coastal Zone Management Advisory Committee, U.S. Department of Commerce, Washington, D.C., June 1978, 10.
66. U.S. Department of Commerce, Office of Coastal Zone Management, CZM Information Exchange, July, 1981, 6.
67. Coastal Zone '78, Symp. Tech., Environ., Socioecon., and Reg. Aspects Coastal Zone Manage., San Francisco, Calif., March 14-16, 1978, Am. Society of Civil Engineers, New York, 2447.
68. Coastal Zone '78. Symp. Tech., Environ., Socioecon., and Reg. Aspects Coastal Zone Manage., San Francisco, Calif., March 14-16, 1978, Am. Society of Civil Engineers, New York, 2467.
69. Coastal States Organization, Coastal Management - Options for the 80's, Fredricksburg, Va., January 1979, 7-12.
70. Coastal States Organization, Coastal Management — Options for the 80's, Fredricksburg, Va., January 1979, 1-4.

71. The First Five Years of Coastal Zone Management, Office of Coastal Zone Management, National Oceanic and Atmospheric Administration, Washington, D.C., March 1979, 60.
72. The First Five Years of Coastal Zone Management, Office of Coastal Zone Management, National Oceanic and Atmospheric Administration, Washington, D.C., March 1979, 1.
73. The First Five Years of Coastal Zone Management, Office of Coastal Zone Management, National Oceanic and Atmospheric Administration, Washington, D.C., March 1979, 2.
74. U.S. Department of Commerce, The Federal Coastal Programs Review - A Report To The President, Washington, D.C., January, 1981, D-4.
75. Problems Continue in the Federal Management of the Coastal Zone Management Program, U.S. General Accounting Office, Washington, D.C., June 1980, 1.
76. Problems Continue in the Federal Management of the Coastal Zone Management Program, U.S. General Accounting Office, Washington, D.C., June 1980, 16.
77. Problems Continue in the Federal Management of the Coastal Zone Management Program, U.S. General Accounting Office, Washington, D.C., June 1980, 15.
78. Coastal Zone Management 1980 - A Context for Debate, The Conservation Foundation, March 1980, 2.
79. Coastal Zone Management 1980 - A Context for Debate, The Conservation Foundation, March 1980, 2.
80. Coastal Zone Management 1980 - A Context for Debate, The Conservation Foundation, March 1980, 3.
81. U.S. Congress, House, Representative Bonoir Speaking for the Coastal Zone Management Improvement Act of 1980, HR 6979, 96th Congress, 2nd Session, September 30, Congressional Record, H 10112, 1981.
82. U.S. Congress, House, A Bill to Improve Coastal Management, HR 6979, 96th Congress, 2nd Session, 1981.
83. Executive Office of the President, Office of Management and Budget, Fiscal Year 1982 Budget Revisions, March, 1981, 49.
84. Legislative History of the Coastal Zone Management Act of 1972, as Amended in 1974 and 1976 with Section-By-Section Index, U.S. Government Printing Office, Washington, D.C., 1976, 191.
85. U.S. Congress, House, A Bill to Improve Coastal Management, HR 6979, 96th Congress, 2nd Session, 1981.
86. California vs. Watt, CV-81-2080 MRP, U.S. District Court, Central District of California, August, 1981, 3-9.
87. California vs. Watt, CV-81-2080 MRP, U.S. District Court, Central District of California, August 1981, 2.
88. California vs. Watt, CV-81-2080 MRP, U.S. District Court, Central District of California, August 1981, 58.
89. California vs. Baldridge, CV-81-3760 CBM, U.S. District Court, Central District of California, July, 1981, 8.
90. California vs. Baldridge, CV-81-3760 CBM, U.S. District Court, Central District of California, July 1981, 1.
91. Coastal Zone Management Newsletter, 12(28), 1. July 22, 1981.
92. Coastal States Organization, Testimony of James F. Ross, Chairman submitted to Subcommittee on Oceanography, Merchant Marine and Fisheries Committee, U.S. House, March 31, 1981.
93. U.S. Department of Commerce, Office of Coastal Zone Management, CZM Information Exchange, July, 1981, 1.
94. U.S. Department of Commerce, The Federal Coastal Programs Review - A Report To The President, Washington, D.C., January, 1981, 1.
95. The Federal Coastal Programs Review — A Report to the President, U.S. Department of Commerce, Washington, D.C., January 1981, 4.
96. U.S. Department of Commerce, Office of Coastal Zone Management, Memo from Robert W. Knecht to Interested Parties transmitting The Federal Coastal Programs Review report, May 13, 1981.
97. Coastal Zone '78, Symp. Tech., Environ., Socioecon., and Reg. Aspects Coastal Zone Manage., San Francisco, Calif., March 14-16, 1978, Am. Society of Civil Engineers, New York, 2468.
98. **Curlin, J. W.,** What the Reagan Bunch Should Know About Ocean Policy . . . And What They Should Do About It, (An address presented to the Marine Technology Society, Fort Meyer, Virginia, January 6, 1981), 2.
99. **Botzum, J. R.,** A National Ocean Policy — Now!, (An address presented to the National Ocean Industries Association, Washington, D.C., March 9, 1981), 1.
100. **Baldridge, M.,** untitled (Remarks delivered at the presentation of the Neptune Award to Senator Ted Stevens, Washington, D.C., April 29, 1981).
101. Coastal State Organization, unpublished survey, March, 1981.

Chapter [illegible]

CALIFORNIA

[illegible]

TABLE OF CONTENTS

Chapter 8

COASTAL LAW

David W. Owens

TABLE OF CONTENTS

I. INTRODUCTION

The law of the coast is made up of thousands of federal, state, and local laws, regulations, and court decisions. The body of law is as complex and dynamic as are coastal resources themselves. Because of this, the resolution of any particular legal issue will vary from state to state and may well change over time. What follows, therefore, is necessarily a general overview of the current state of coastal law in the U.S. rather than a detailed examination of the law of any one particular jurisdiction.

This chapter examines the issues of who owns various coastal resources and the character of that ownership. It addresses major trends in a governmental regulation of coastal resources at the federal, state, and local level, giving particular attention to the purposes and mechanics of regulation. The constitutional issues raised by these regulations are also discussed.

II. OWNERSHIP AND USE

A. Basic Boundaries and Underlying Ownership

Land ownership and property lines are usually clear and stable in upland areas. Absent extraordinary circumstances, it is not difficult to survey precisely where ownership boundaries are located or to determine specifically who owns a given piece of property. The process may be expensive, involving lawyers and surveyors, but the answers are usually certain and the law is clear. This is not the case with coastal properties. Here, legal scholars have debated for centuries the questions of who owns beaches and wetlands, what public rights exist in these areas, and even how to determine boundaries.[1-7]

The physical points generally used to define where property rights change are shown in Figure 1. On the oceanfront and the shores of the Great Lakes, the key dividing lines are the mean low water line, the mean high water line, and the vegetation line. Over time these lines have come to have precise legal definitions, definitions that vary from state to state.

Ascertaining where a mean tide line is physically located is a complex task, both technically and legally.[8-13] Most technical definitions are based on determining a vertical tidal datum and plotting the intersection of that datum plane with the shoreline. For example, to determine the mean high tide line, the average elevation of the sea level at a normal high tide period (measured over an almost 19-year period in order to account for the cycles the moon goes through in relation to the earth) is established. Then the point at which this vertical datum plan intersects the beach (e.g., the 4.6 ft contour) is technically the "mean high water line". Since the beach profile is constantly changing with daily and seasonal cycles, this line is constantly moving. Not all states follow this technical method of establishing property lines. Wisconsin, for example, has long used physical indicators (trash line, erosion scarp, etc.) to establish the high water mark.

Because of the difficulty in precisely establishing such lines and this constant movement, there are suggestions increasingly being made to use a more stable line as a point of reference, such as the vegetation line. In Hawaii the vegetation line is generally used now because pre-statehood law used the "upper reaches of the wash of the waves" as the key reference point and modern court decisions have ruled this to be the vegetation line.[14,15] Most states do not use the vegetation line, however, as a reference point for ownership, though many now use it as a reference point in regulations.

In a majority of oceanfront states, the mean high water line is the boundary between public and private ownership. The dry-sand beach above the mean high water line is privately owned while the wet-sand beach below this line is owned by the state. These "high-tide" states include Alabama, Alaska, California, Florida, Georgia, Louisiana, Maryland, Mississippi, New Jersey, New York, North Carolina, Oregon, Rhode Island, South Carolina, Texas, and Washington. A few states, primarily in the northeastern part of the country, hold

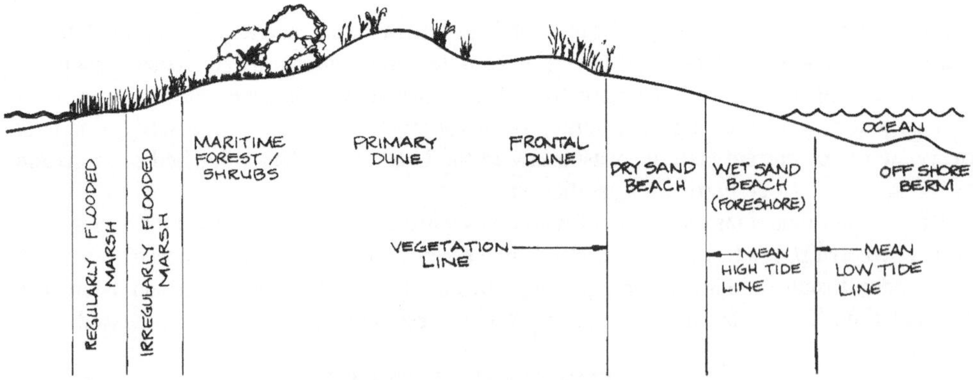

FIGURE 1. Typical barrier island profile. (Drawn by J. Miller.)

that private ownership runs to the mean low water line, thereby making the wet-sand beach privately owned. These states include Delaware, Maine, Massachusetts, New Hampshire, Pennsylvania, and Virginia.

The situation regarding the ownership of wetlands is just as complex.[16-23] While the general rule of the mean high water line separating public and private ownership usually holds, the situation is complicated by the fact that some wetlands are regularly flooded (by daily tides) while others are only irregularly flooded (by wind tides, storm tides, etc.). So part of the "wetlands" may be publicly owned, part privately owned. Also, ownership may be affected by the questions of whether the wetlands are adjacent to "navigable" waters.[17]

If the waters are "navigable" the public usually has a number of rights in those waters and their bottoms, such as ownership of the bottoms, public rights of navigation and fishing, and title to adjacent regularly flooded marshes. Each state has evolved its own test of navigability, with the more common tests being whether the water is subject to the ebb and flow of the tides, whether it is navigable by commercial vessels, and whether it is navigable by any boats commonly used in the area, including shallow draft boats such as skiffs and canoes.

B. Character of Public Ownership: The Public Trust Doctrine

Those coastal lands that are owned by the public are held in a special capacity as they are held by the state in trust for all the people of the state. This public trust doctrine has a long legal history.[24-26] In Roman law the sea and its shores were held to be *res communes*, open to the use and enjoyment of all citizens. Similarly major navigable rivers and their banks were held to be *res publicae*, publicly owned. The state's ownership of such areas was not proprietary or exclusive. Rather than state owned the area in a trustee capacity, protecting all of the citizens' rights of navigation and fishing.

The English common law recognized a similar public trust doctrine. Those areas subject to tidal action were owned by the Crown as *jus publicum*, subject to public rights of fishing and navigation. The King owned these lands as the representative of the nation and for the public benefit. These lands could be sold by the Crown to individuals if expressly granted, but the lands remained subject to the public's rights.

Following the American Revolution, the former colonists generally adopted the English common law as the basic law of the new country and the public title to these lands vested in the various states. The character of this ownership was early on held to be of the same nature of the Crown's prior ownership, *jus publicum*.[27-30] As such, these lands were held in public trust for the common use of all.

It was not until 1892, however, that the U.S. Supreme Court established detailed guidelines for the states' responsibility under the public trust doctrine.[30] This case involved an attempt by the 1869 Illinois legislature to grant an extensive area of the Chicago waterfront and the adjacent submerged area of Lake Michigan to the Illinois Central Railroad Company. The 1873 legislature sought to invalidate this grant and litigation ensued. When the case finally reached the U.S. Supreme Court, it held that the state owned the lands beneath Lake Michigan in the same capacity that states owned lands covered by tidal waters. The court went on to hold that such lands are "held in trust for the people of the state that they may enjoy the navigation of the waters, carry on commerce over them, and have liberty of fishing therein freed from the obstruction or interference of private parties." The court held that given this public trust, only those grants made to promote these public interests and which do not substantially impair such interests can be allowed. The court therefore ruled that the 1869 grant to the railroad company was ineffective because the state will not be allowed to abdicate its public trust responsibilities.

Two of the key issues concerning the public trust doctrine are: (1) to what lands and waters does it apply; and (2) what public interests are protected by the doctrine. Courts have for the most part limited application of the doctrine to tidal and navigable waters up to the mean high water line.[8,9] The public rights protected traditionally included navigation, commerce, and fisheries, with some modern decisions including the right to hunt, swim, and use the area for general recreational purposes.[31] Private littoral or riparian owners have generally been held to have the right to "wharf out", to build a pier extending to navigable waters, and to have rights of access across all of their water frontage.

Another important question is what impact dynamic physical features have on public and private property rights.[32] Since mean tide lines are used as property lines, what happens when the tide line moves? Given the highly dynamic nature of these lines (with erosion, accretion, reliction, and avulsion) the normal certainty of stable porperty lines does not hold. The general rule of law is that if the physical change is gradual, as with long-term erosion, accretion or reliction, then the property line also moves. That is, land lost to gradual erosion belongs to the state and land gained by gradual accretion belongs to the littoral/riparian owner. The converse of this rule is that if the change is sudden and perceptible, the pre-existing property lines are not affected.

Given the certain seasonal fluctuations of the ocean shoreline, at least one case has held that this normal ambulatory property line concept should not be followed in ocean-front situations. A Florida lower court held that the furthest seasonal landward extent of the mean high water line (which occurs in winter when the beaches are naturally narrower) should be used as the property line. [33] Although criticized by some commentators, this rule may more accurately reflect physical, cultural and social realities than use of the traditional ambulatory mean high tide line concept.

C. Public Use Rights in Private Lands

Even though public ownership may end at the mean high water line, a number of state courts have held that the public has certain rights in adjacent dry-sand beach areas.[34-36] As more and more people sought to use scarce beaches for recreation in the 1960s and 1970s, clashes with beach land owners over who could use the beaches became commonplace across the nation. A number of key states, using different legal doctrines, held that even though the beach area was privately owned, the public could not be prevented from using it for recreation and access.

One of the more interesting and controversial legal doctrines in this field is that of "customary rights"; an old English common law theory used by the Oregon Supreme Court in a beach use case in 1969.[37] In this case, a motel owner had in 1966 fenced in the dry-sand portion of the beach adjacent to his motel for the exclusive use of motel guests. The state's legislature then passed a law recognizing the public's interest in open beaches and requiring permits for future fences of this sort. When the motel's fence was destroyed in a storm and rebuilt without a permit, the state took the motel owner to court to have the fence removed. Testimony at the trial showed that, prior to erection of the fence, the public had used this beach area for over 60 years without ever securing the landowner's permission. Because of the nature of ocean beaches and in order to avoid numerous individual legal disputes, the court held that such long standing use had given the public a "customary right" to continued use of the state's similar dry-sand beaches.

The seven legal requirements for establishing "customary rights" at common law are as follows:

1. The use must be "ancient," so long that nobody remembers otherwise.
2. The use must have been without interruption, which is to say the public must not have been excluded by the upland owners during the period.
3. The use must have been peaceable and free from dispute, the public entering and using the area without resort to force.
4. The use must have been reasonable and in keeping with the character of the land.
5. There must be certainty as to just what land was being used.
6. The use must have been obligatory for the upland landowners, that is, the public's use not being subject to the option of each individual upland owner.
7. The use must not be repugnant or inconsistent with public policy and other laws.

No other state courts have yet adopted this doctrine in a coastal law context and some commentators have criticized it as being overly broad.[38,39] However, in situations where the public and the littoral landowners have always assumed that the public had rights to use the dry-sand beach below the vegetation line, application of such a doctrine seems equitable and consistent with sound public policy.

A second legal doctrine that has been used to establish public use and access rights in privately owned beach areas in the "implied dedication" concept. The leading case using the implied dedication theory is a 1970 California Supreme Court decision.[40] This decision provoked, more so than any other single court decision on beach access, tremendous controversy and discussion among legal scholars.[38,41-49] The doctrine has also been used, at least in part, by the Texas court.[50]

The situations in these decisions involved both public use of the dry-sand beach for recreation and use by the public of accessways across upland property to get to the beach. The court held that, based on long standing public use, the law would imply that the owners had dedicated an easement to the public for continued use.

This legal doctrine of "implied dedication" does not depend on the owner having made a written or express grant to the public for use of the property. Rather then owner's behavior is not effectively excluding public use leads the law to imply that a dedication was intended, even though the owner's actual intent may have been quite different. The public's acceptance of the dedication is also implied by their conduct; the use of the beach over an extended period without seeking the owner's permission. Once the requirements for implied dedication have been met, the dedication to the public is complete and irrevocable. Future lack of use by the public or exclusion of the public by the owner does not cause the legal use rights to be lost.

A legal doctrine that is very closely related to that of "implied dedications" is the "prescriptive easement" concept. This doctrine has been used in Florida[51,52] and Texas[50] beach use cases. The public may obtain a legal right, an easement, to continued use of a beach if the tests for a prescriptive easement are met. While the exact test varies form state to state, the general requirements are that there be

1. Actual use of the privately owned property by the public
2. Continuous and uninterrupted use for a set period of time (often seven to twenty years)
3. Open and clearly visible use (so as to give notice to the owner)
4. A claim of right to use the property by the user
5. Use in a manner that is adverse to the interests of the owner

This legal concept's use in a beach use/access context is limited in several respects. Proving continuous use over a long period of time at a trial can be difficult and costly. The issue generally has to be litigated on a parcel by parcel basis. Also, a number of courts will hold that public use of an open, unenclosed beach area is by the implied permission of the landowner and not adverse to the owner's property interests.

A legal issue similar to these questions of public use of privately owned coastal beaches also arises with limitations imposed on use of publicly owned beach areas. In a number of beach communities part of the beach front is owned by the local government as a park. Some communities, particularly in the northeast, have attempted to prevent nonlocal residents from using their beach parks or have tried to charge nonresidents significantly higher use fees.[53,54] The New Jersey and New York courts have invalidated such local ordinances. The New Jersey case involved discriminatory user fees. The state supreme court ruled that the public trust doctrine applies to municipally owned dry-sand beaches and that they must therefore be open to all members of the public on a nondiscriminatory basis.[55,56] The New York case involved an attempt to totally exclude nonresidents from the local beach. There the state court ruled that in acquiring the beach, using federal funds to improve it, and having it open to the general public for over three decades, there had been an express dedication of the park to use by the general public, including nonresidents.[57] Once such a dedication has been made, it is irrevocable.

The question of public access to dry-sand beaches remains a controversial issue.[58-62] While a number of state courts have confirmed public rights in these areas, several states have refused to follow the trend.[63-66] Other key states have not yet resolved these difficult legal questions. One of the few certainties on this particular issue is that it will remain controversial and therefore continue to appear before the courts and legislatures.

III. LEGAL ISSUES IN COASTAL MANAGEMENT AND REGULATION

A. Purposes of Management

Before examining the legal questions raised by governmental management of coastal resources, it is helpful to briefly examine the purposes behind these programs. Justifications most frequently offered are minimizing losses to life and property in natural hazard areas and protecting important natural functions from undue harmful human interference.

A great deal of attention has recently been focused on the hazard mitigation purpose.[67,68] Given the highly dynamic and sometimes violent nature of coastal resources, such attention is warranted. Most of the open coast is subject to periodic major storm events; hurricanes, typhoons, and extratropical depressions. When these occur, high winds, high tides, flooding, storm surge, and violent wave action can combine to cause drastic impacts, at least temporarily, on both natural and man-made features. Whole dune fields are inundated and pulled off-shore to become underwater sand bars. Coastal shorelines may recede several hundred feet. New inlets may be created and existing inlets dramatically expanded and altered. Heavy rains can cause severe flooding of low-lying areas. In addition to these storm hazards, there are other unique coastal hazards, such as migrating inlets and long-term gradual erosion resulting from sea level rise.[69,70]

The public costs resulting from the damage caused by these natural hazards can be quite large. After a storm tremendous sums are spent for disaster relief, addressing both short-term needs (e.g., evacuation, medical assistance, and temporary housing) and the long-term costs of rebuilding (e.g., low interest loans to restablish businesses.). Private property losses are also very high. Estimates of losses due to 1979s Hurricane Frederick range as high as $2 billion. Because the risk of catastrophic loss is too high for private insurance companies, the federal government has since the early 1970s provided subsidized flood insurance for these areas at a total national cost to the taxpayers of $850 million over premiums received for the 1970s alone.

Preventing storm losses has been the justification for additional millions of dollars to be spent on public erosion control projects, ranging from seawalls in Texas to huge boulder rip-rap in New Jersey, artificial sand berms in North Carolina and artificial beach nourishment in Florida. Most of these large scale projects have been financed primarily by the federal government, with states and local governments funding lesser amounts.

Another significant public cost has been the repair and replacement of those public investments damaged by storms, such as roads, bridges, and water and sewer lines. In the two years after Hurricane Frederick, about $45 million of public funds were spent in restoration efforts on Dauphin Island, Alabama alone, a small resort island with only 650 permanent structures when the storm hit. A legal challenge by environmental groups to the replacement of a $38 million bridge to this island, which experienced complete overwash in the storm, was unsuccessful.[71,72]

A second major purpose of governmental regulation of coastal resources is the protection of their natural values. Coastal waters and wetlands are critical for commercial and recreation fisheries. Wetlands in a natural state can be important for a number of reasons. They protect water quality by serving as a natural nutrient and sediment trap. They protect fish and wildlife populations by serving as a spawning and habitat area and contributing detritus to the food chain. They provide drinking water by serving as groundwater recharge areas. Wetlands also protect development from natural hazards, serving as a buffer to erosion and as a storage area for flood waters. A number of other coastal locations are habitat for rare and endangered plant and animal species. Sand dunes can provide some protection for development, particularly in modest storms. In addition, the simple natural beauty of coastal areas is a principal ingredient in their attraction to both residents and to tourists. Maintaining these natural values therefore has important practical and economic benefits.

B. Principal Management Tools

1. Federal Level

There are a number of programs carried out by federal, state, and local governments designed to manage coastal resources. Each raises unique legal questions regarding its scope, purpose, and application.

Several federal laws set direct standards for development in coastal areas. The U.S. Army Corps of Engineers administers two of the most important of these; "Section 10" and "Section 404" permits. Section 10 of the U.S. Rivers and Harbors Act of 1899 requires a permit from the Corps for any obstruction or alteration of navigable waters. This includes both structures, such as docks and piers, and physical alterations such as dredging and filling. Section 404 of the Federal Water Pollution Control Act Amendments of 1972 requires a Corps permit for the discharge of dredged or fill material in any wetland areas. Under these laws, the Corps is charged with making careful reviews to assure the proposed activity is in the public interest prior to granting approval, with environmental factors being a key element of the review.[73-75] The Section 404 permit is particularly important because it applies to all wetland areas, not just navigable waters. In fact, the courts have held that the government's interest and authority to regulate can extend to privately-owned water bodies.[76,77]

Given this broad geographic scope, the broad "public interest" reviews used to decide permit applications, and the presumption used by the Corps that wetlands should generally remain in their natural state, this law is in many respects the central federal regulatory mechanism for managing coastal resources.

There are several other federal regulatory provisions that affect coastal resource management. The Marine Protection, Research, and Sanctuaries Act of 1972 regulates the dumping of materials in the ocean. The Clean Air Act of 1970 regulates the emission of air pollutants, preventing new pollution sources that would violate established emission standards or that would lead to a significant deterioration of air quality. The Federal Water Pollution Control Act amendments of 1972 establishes the National Pollution Discharge Elimination System (NPDES). This Act requires a permit for the discharge of pollutants from any point source (e.g., a pipe, ditch or other discrete or confined source), with detailed standards being set for the content of discharges. These permits are generally administered by the states according to national standards set by the U.S. Environmental Protection Agency.

Several other federal laws rely on public disclosure of environmental information rather than direct regulation. The most important of these is the National Environmental Policy Act of 1969 (NEPA).[78-83] This Act requires federal agencies to prepare an environmental impact statement (EIS) before making major decisions that have a significant effect on the environment. The decisions covered include activities such as decision on major permit applications, major construction projects, and large grants. In all instances, the NEPA process is triggered if there will be a significant effect on the quality of the environment as a result of the decision. The required statements must address the environmental impact of the proposed decision, adverse environmental impacts that cannot be avoided, alternative decisions considered, the relationship between short-term use of resources and their long-term productivity, and any irreversible or irretrievable resource comments involved. The principal purpose of these statements is to improve the governmental decision-making processes by requiring information and analysis on the environmental impacts of decisions to be gathered and presented to the decision-makers prior to their decision. To this end, the thick technical documents that characterized early EIS's have in recent years been replaced with more concise documents that can more practically be used by agency decision-makers. Findings that there will be significant adverse environmental impacts does not necessarily mean the project can not go forward, however, as NEPA only requires that the environmental effects be considered prior to reaching a decision.[84] Other major purposes of NEPA include: assuring that environmental factors are considered by all agencies, even those single-purpose agencies who previously considered environmental concerns outside their sphere of interest; improving interagency and intergovernmental coordination through circulation and review of impact statements; and disclosure to the public of the potential impacts of decisions being considered.

Another major "disclosure" law is the Interstate Land Sales Act of 1969. Where lots in subdivision of more than 50 lots are being offered to buyers in more than one state, as is frequently the case with coastal developments, the developer must register with the federal government, prepare a detailed analysis of the subdivision (noting, e.g., flood hazards and actual improvements in place) and make the statement available to all prospective purchasers prior to sales.

There are several federal laws that provide incentives to state and local governments to manage development of coastal resources. Two of the most important of those are the flood insurance program and the coastal zone management program.

The National Flood Insurance Act of 1968 establishes a program whereby the federal government will subsidize insurance for structures and their contents against flood and storm loss if the local community is engaged in managing development in flood hazard areas. Only losses due to storm events, as opposed to long-term gradual erosion, are covered. The subsidized insurance is offered in two steps. The first is the "emergency phase".

Communities are eligible for this phase after a generalized flood hazard map has been prepared and maximum coverage limits are 1/2 those in the regular phase. Communities can move into the "regular phase," thereby doubling the insurance coverage available, only after detailed flood insurance rate maps and local floodplain management regulations have been adopted. The rate maps detail various risk zones, delineating areas subject to flooding and areas subject to wave action as well as flooding in different intensity storms. The maps are used in part to set insurance premiums and in part to define areas of regulatory coverage. The local regulations that are adopted must meet federal standards designed to minimize future flood losses. The minimum standards include provisions such as elevating the ground floor of structures above the 100 year storm level, designing utility systems so as to prevent infiltration of storm waters, and designing foundations to withstand normal storm conditions. Congress in 1981 passed legislation to prohibit issuance of new flood insurance policies on undeveloped barrier beaches, effective October 1, 1983.

The federal Coastal Zone Management Act of 1972 (FCZMA) establishes several strong incentives for state governments to develop and improve state and local coastal management programs. The principal incentive was the availability of federal funds to finance 80% of the cost of developing and implementing state programs that meet federal standards. To be eligible for funding and federal approval, state programs must, among other requirements, include:

1. An identification of the boundary of the state's coastal zone
2. A definition of those permissible land and water uses within the coastal zone that have a direct and significant impact on coastal waters
3. An explanation of how these uses will be managed
4. An inventory and designation of areas of particular concern and the ability to designate areas for preservation and restoration
5. General priorities of uses in these particular areas
6. A description of the program's organization
7. A planning program for access to beaches and other important public coastal areas
8. A planning program for siting energy facilities in the coastal area
9. A planning program for addressing shoreline erosion
10. A program for coordinating program development and implementation with all affected parties, public and private.
11. A demonstration that the program provides for adequate consideration of the national interest in planning and siting facilities serving other than local needs
12. A method for assuring that local regulations do not unreasonably restrict or exclude land and water uses of regional benefit

The management program can be carried out directly by the state government or parts can be handled by local governments, provided there is adequate state oversight. The FCZMA also provides for a grant program to fund land acquisition and construction of facilities for preserving natural areas, revitalizing urban waterfronts, and providing public beach access, although these sections of the FCZMA have never been funded.

The only significant legal issues to arise with these provisions of the FCZMA involved the federal government's decision in 1978 to approve state programs developed by the states of California, Massachusetts, and Wisconsin. The challenges were brought by the oil industry and alleged the programs did not meet federal approval standards. The suits on the Massachusetts and Wisconsin approvals were dismissed when the federal courts found the challenges premature (injuries to the oil industry were merely speculative) and the companies lacked standing to sue (they had no legally protected interests being violated by the approval).

In the California challenge, the court upheld the federal approval decision, holding the program met all approval standards.

The other incentive for states to develop coastal management programs provided by the FCZMA was its requirement that once a state has an approved program, most federal activities are required, to the maximum extent practicable, and to be consistent with the state program. This requirement applies to all federal actions that "directly affect" a state's coastal zone, including federal permit decisions, federal grant decisions, and direct federal development activity. The most significant legal controversy to date in the interpretation of this provision of the FCZMA has been in determining when a federal decision will "directly affect" a state's resources. Federal agencies have contended that with Outer Continental Shelf oil and gas development, direct effects arise only at the exploration and development stages, not at the off-shore leasing stage. The coastal states have contended that the lease decisions themselves directly affect their coasts and coastal waters and that the consistency provision of the FCZMA should apply at this stage also. While the issue is still being litigated, to date the courts have accepted the state position.

There are a number of other federal laws that affect coastal resource management though they are not management tools per se. Key among these are federal investment and acquisition programs. The federal government can and has acquired critical coastal natural areas as wildlife refuges, national seashores, and national parks. Other areas have been designated marine and estuarine sanctuaries. Federal investments provide utility systems that support development and hurricane protection/erosion control projects to protect coastal development. The way these programs are administered has a great deal of impact on effective coastal resource management. While at this time there are few attempts to explicitly use these programs to further coastal management objectives, Congress has adopted legislation, the Coastal Barrier Resources Act of 1982, to limit new federal investments on undeveloped barrier islands.[85]

2. State and Local Level

Coastal management and law at the state and local level varies tremendously from state to state. There are very active management programs at the state level in some states, while in others there is only rudimentary state level management. Local management efforts also vary considerably, not only nationally but even within individual states.

While coastal local governments have a long history of management efforts, the state's role in coastal management is more recent, generally dating only from the 1960s. In response to a growing awareness of the necessity of proper management of the environment, a recognition that many coastal issues transcend local boundaries, and increasing pressures on sensitive coastal resources, many states moved to enact state coastal management laws.[86-103] Financial incentives for adoption and implementation of state management programs provided by the federal government provided an important impetus to this movement in the 1970s. Some laws established direct state regulation of development of coastal resources, many established state-local management partnerships, and others mandated a primarily local management program with some state oversight. By 1981, 26 of the 35 coastal states and territories had coastal management programs that had been approved by the federal government as meeting the standards set for funding under the Federal Coastal Zone Management Act.

One of the key legal issues regarding state coastal management efforts has been the degree to which state legislatures can delegate management decisions to specialized administrative boards, commissions or agencies. Since coastal management decisions are difficult, involving complex natural resources and a balancing of many competing interests, those preparing coastal legislation found it impossible to establish precise standards and decision guides in the legislation setting up state coastal programs. Therefore several legislatures created specialized administrative bodies to carry out coastal laws. These bodies have been empowered to set the geographic limits of permit jurisdiction, establish the standards for coastal development, and to make permit decisions, grant variances, and resolve administrative appeals of agency decisions, all within broad guidelines established by the legislature.

There is a fundamental rule of American law requiring the separation of power among the legislative, executive, and judicial branches of government. If the courts find that a statute delegates basic legislative policy-making authority to executive boards and agencies, the statute may be invalided as an unlawful delegation of legislative authority. However, because of the complexity of modern governmental decision-making, there is a trend in the judiciary to approve very broad delegations of authority so long as there are sufficient general legislative standards to guide implementation of the law and adequate procedural safeguards to assure that the general legislative intent is being followed. Coastal legislation making such delegations to boards and commissions in California,[104] New Jersey,[105] and North Carolina[106] has been found by the courts in those states to be valid in the face of challenges regarding the extent of delegation. However, statewide legislation in Florida allowing an administrative commission to designate "areas of critical state concern" and approve management plans for the areas so designated was held to be an invalid delegation of legislative authority.[107] While in many respects the Florida legislation made no broader a delegation than that approved in other states, the Florida court was concerned about the absence of legislative priorities for which resources should be protected and the fact that the law was not confined to a defined high priority part of the state, as was done with coastal legislation in California, New Jersey, and North Carolina.

Limiting a state law to application only in part of a state, such as the coastal zone, raises another legal issue. Many states prohibit "local legislation" on topics of statewide concern. This was a major aspect of the challenge to the legality of North Carolina's coastal legislation.[106] There the court ruled that application of the law to just coastal counties was valid given the unique, significant, and urgent environmental problems found in the coastal zone.

There are a number of management tools used by state and local governments to manage coastal development. One survey indicated that of those 26 states and territories with federally approved coastal management programs, the principal areas of management attention are as follows: 26 have wetland, beach, and dune protection programs; 24 have management of erosion-prone and flood plain areas; 15 have dredge spoil disposal restrictions; and 14 have energy facility siting provisions.[108] Another survey of local governments participating in the National Flood Insurance Program indicates use of a variety of management tools at the local level to address such concerns.[109] Of those communities in the regular phase of the program, over 75% use elevation requirements, zoning and subdivision regulations, over half use flood proofing requirements, floodway regulations, and comprehensive planning, and over a quarter of the coastal communities also use septic tank, sedimentation and erosion control, wetland, and sand dune protection regulations acquisition of open space, and information programs on hazard areas.

In managing hazard areas, one of the most direct management tools used is requiring a minimum setback of structures from the ocean or other natural hazard.[110,111] Such setbacks can be based on past erosion rates, predicted future erosion or storm recession, the presence of protective sand dunes, or a set distance that considers these and other factors.Other hazard area regulations that are frequently used include limits on the alteration of frontal or primary sand dunes, prohibition of the destruction of dune vegetation, and minimum construction standards to assure elevation of structures, adequate foundation support (particularly respecting pile size and depth) and adequate roof design and tie-downs to minimize wind damage. As for protection of natural areas, the coastal regulations most frequently used at the state and local level are prohibitions of or limitations on dredging and filling in salt marshes and other wetland areas, limitations on the placement of structures in navigable waters, and special rules on development in critical areas, such as primary fishery nursery areas or endangered species habitats.

State and local governments also make extensive use of nonregulatory coastal management efforts. Education as to coastal hazards and natural values is frequently used. This can be done on a general level or can be required on an individual site by site basis. For example, the courts have upheld a local requirement for a professional geologic soils report prior to building approval.[112] Also, the Oregon courts have held, in a case brought by owners of oceanfront condominiums being destroyed by erosion, that the developer/seller of coastal real estate has a duty to exercise reasonable care to inspect the land for its suitability for the intended use.[113] Findings of such a duty help lessen the inequities created by a strict adherence to the traditional "buyer beware" rule, as many coastal hazards are not readily apparent to the nonprofessional.

C. Constitutional Issues

All governmental efforts to manage coastal resources must be carried out within the basic legal framework established by the U.S. Constitution and similar state constitutions. If the courts find that a law or ordinance violates a constitutional requirement, the offending law will generally be invalidated, though in some limited circumstances the remedy provided by the courts will be compensation for the effected landowner.[114] The principal constitutional law issue affecting coastal resource management is the prohibition against the taking of private property by the public without just compensation. The guarantee that the due process of the law will be followed in all governmental actions and other constitutional issues, such as the equal protection clause, the right to travel, and various state constitutional provisions regarding rights to environmental quality, may also be of concern in specialized situations.[115]

1. The Takings Clause

The Fifth Amendment to the Constitution of the U.S. provides ". . . nor shall private property be taken for public use without just compensation." This applies to all levels of government; federal, state, and local.

This prohibition of government seizure of private lands without a fair payment being made to the private owner came to America directly from the English common law. As the law of property began to develop in England, the Crown owned all lands. Nobles, in return for favors provided to the Crown, such as provision of troops, payment of a fee, or the like, were given actual possession of lands. The nobles in turn were allowed to pass possession along to others in return for favors provided to the nobles. Under this feudal system, any failure to provide the favors when due resulted in a forfeiture of the right of possession of the property to the person to whom the favor was due. As these favors became more burdensome, due to unsuccessful military campaigns, the number of forfeitures rose. As the forfeitures rose, so did the level of dissent between the nobles and the Crown. This led, in 1215, to the inclusion in Magna Carta of the forerunner of the "takings" clause, which provided that "No freeman shall be . . . deprived of his freehold . . . unless by the lawful judgement of his peers and by the law of the land." Over the following centuries, it became established as a fundamental limitation on the power of the Crown that properties granted to private parties could not be arbitrarily reclaimed by the Crown. If the property was seized for public use by the Crown, fair compensation had to be provided.

This general concept was incorporated into American law in the earliest days of the republic. Inclusion of the "takings clause" into the Bill of Rights assured that the federal government would be subject to this limitation, just as the Crown was in England. While this originally applied only to the federal government, by the end of the 19th century the courts had ruled that the Fourteenth Amendment, which provides "nor shall any state deprive any person of life, liberty, or property without due process of law," extended this limitation to state and local governments.

It is clear that when government actually takes title to property, as for example in acquiring land for a post office or a park, compensation must be paid. The more difficult question is determining what other governmental actions require compensation.

Early judicial interpretations of this constitutional limit involved actual physical invasions of private property. For example, the Supreme Court ruled in 1871 that where the state authorizes a dam and floods private property, that property has been "taken" and compensation must be paid to the property owners.[116] This type of physical invasion can arise in a coastal context. For example, a New Jersey town decided to construct a protective sand dune along the oceanfront to minimize future storm damage to public and private properties. When the dune was built on private property without permission or compensation, the New Jersey courts held this to be an unconstitutional "taking" of the private property.[117]

However, the principal modern application of the takings clause to coastal management concerns the question of whether regulations that greatly restrict the use of private property amount to the equivalent of a governmental "taking" of the property. Early Supreme Court decisions indicated that regulations adopted under the police power to protect the public welfare would not be considered a "taking". For example, the court ruled in 1887 that a Kansas law prohibiting the manufacture and sale of intoxicants did not constitute a constructive taking of a privately owned brewery, even though the property was rendered useless by the law.[118] The court ruled that "A prohibition simply upon the use of property for purposes that are declared, by valid legislation, to be injurious to the health, morals or safety of the community cannot, in any just sense, be deemed a taking or an appropriation of property for the public benefit." This ruling would indicate that regulations to protect these public interests would never be deemed a "taking". However, the Supreme Court reached a different interpretation in ruling a Pennsylvania antisubsidence law invalid.[119] The state had passed a law that prohibited coal mining where it was likely that the mining would cause harmful subsidence. In an opinion by Justice Holmes, the court in 1922 stated, "The general rule at least is, that while property may be regulated to a certain extent, if regulation goes too far it will be recognized as a taking." Since the court found the regulation made coal mining commercially impractical and that this had the same general effect as appropriating or destroying the property rights of the coal company, the statute was invalidated. In this respect the court admonished, "We are in danger of forgetting that a strong public desire to improve the public condition is not enough to warrant achieving the desire by a shorter cut than the constitutional way of paying for the changes."

Subsequent Supreme Court decisions have not established a clear line of when a regulation will have been deemed to have gone "too far." For example, the court in 1928 upheld a Virginia statute that required the destruction without compensation of privately owned trees to prevent the spread of cedar rust disease.[120] The court also ruled in 1928 that a local zoning ordinance restricting land to a residential use is invalid if it does not promote the health, safety, convenience or general welfare of the city.[121] But the delineation of a precise line between a permissible regulation and an unconstitutional "taking" has long defied precise definition by the courts or by legal scholars.[122-129]

A number of tests have been used by the courts in analyzing when a regulation constitutes a "taking." Perhaps the most common route of analysis has been to ask whether the regulation is designed to obtain a previously unsecured public benefit or to prevent a use that is injurious to others, with the former being prohibited and the later allowed. This analysis focuses on the objective of the regulation and generally upholds "nuisance prevention" types of regulations. Another line of analysis focuses not on the objective of the regulation but on the effect of the regulation on the landowner. Under this approach the courts examine the extent of the diminution of value of the property, asking whether the owner is left with any practical or economically viable uses of the property. If not, the regulation is invalidated. In these cases the courts seem to be primarily concerned with the question of whether individual property owners are being unfairly singled out to bear costs that should be shared by the general public. A third line of analysis combines the above two tests into a balancing test that examines the extent of public gain relative to private harm. A regulation conferring a large benefit to important public objectives with modest private losses is allowed, while one providing modest public benefits with severe private losses is not. Defining exactly where this balance lies is then resolved on a case by case basis.

The application of these principles to coastal management cases, both hazard area management and natural area protection, has led to mixed results.

As for coastal hazard area management, "taking" challenges have arisen regarding state and local regulations preventing construction of structures on frontal dunes or too close to the ocean. While setbacks from streets have long been employed by local governments and approved by the courts,[130] oceanfront setbacks have only recently been adopted on a wide scale basis, so there have been only a few tests of their constitutionality. In one early test, a lower court in New York ruled in 1955 that a local ordinance requiring a 100-ft oceanfront setback was a "takings" and was invalid.[131] This ruling was followed by a lower court in New York in 1977 when it ruled that a local sand dune protection ordinance prohibiting development on an 80-ft deep oceanfront lot constituted a "taking".[132] However, courts outside of New York reviewing similar regulations have found oceanfront setback requirements valid.

The New Jersey court ruled that a local ordinance prohibiting building in the frontal dune area was valid as long as structures could not safely be built on the lots.[133] The Maine court upheld a town ordinance prohibiting location of structures within 75 ft of the mean high water mark.[134] And a Florida lower court has upheld a local ordinance requiring a setback of 25 ft from the erosion scarp or 50 ft from the mean high water line, whichever is the greater distance landward.[135] While there is no clear line of analysis common to each of these cases, it appears a number of courts will be willing to uphold regulations prohibiting location of permanent structures on hazardous oceanfront lots, provided there is adequate evidence showing that: (1) improper development of such lots can harm neighboring properties and natural features; (2) there are important public benefits to be met by proper management, such as protecting public safety in storms and reducing disaster relief, infrastructure repair, and other public costs; and (3) that practical uses of the property that are compatible with its natural character can be carried out, such as beach access, camping or recreation. Still, some courts may find the burden on landowner to be too onerous and may therefore rule such regulations a "taking". Detailed state by state analysis[136] and litigation are needed to fully resolve this question.

As for natural area management, the "takings" challenge has most frequently arisen with regard to regulations strictly limiting a landowner's right to dredge or fill wetlands. Perhaps the most famous case on this topic is a 1972 Wisconsin decision upholding the validity of that state's shoreland zoning law.[137] A test of the law arose when a land owner filled wetlands along the shoreline without attempting to secure a necessary permit. The court ruled the law valid, holding that "an owner of land has no absolute and unlimited right to change the essential character of his land so as to use it for a purpose for which it was unsuited in its natural state and which injures the rights of others". The important natural values of wetlands and their integral relationship to navigable waters subject to the public trust were important factors to the court in reaching this judgment. In a somewhat similar ruling, the Florida court upheld a state decision to prohibit the alteration of 1,800 acres of black mangroves in an overall development project of 6,500 acres.[138] The court ruled that protection of sensitive environmental resources and prevention of pollution were legitimate objectives of government regulation. The Florida court agreed with the Wisconsin court that a regulation of this type prevented a potentially significant harm to the public and therefore ruled that there was no "taking".

2. Other Constitutional Challenges

Of the other constitutional issues,[115] the most significant in a coastal management context is the due process clause. In this respect the courts will examine whether a regulation is "fundamentally fair." Topics addressed in this analysis include questions such as whether the regulation includes adequate standards to prevent arbitrary application and whether fair procedures are followed, such as giving affected parties a reasonable chance to be heard and having an impartial tribunal to hear appeals. With "legislative" decisions (such as adopting a regulation) the courts will generally make a presumption of validity of the legislative act. If the regulation bears a rational relationship to the accomplishment of a legitimate governmental objective, the courts will generally uphold it. More vigorous review is given to "administrative" decisions, such as the hearing of an individual permit appeal. In either case debatable policy choices are left by the court to legislative judgment. Another constitutional provision that may arise is the equal protection clause. In this context, the courts will invalidate regulations that treat similarly situated persons differently. The judicial review on this point is very stringent if persons are being treated differently on the basis of race. The equal protection clause, however, does not require application of the same rule to all persons. If persons are not "similarly situated," different rules can apply. For example, a rule can be applied prospectively to only apply to property proposed to be developed after the effective date of the rule or there can be different rules for developed and undeveloped areas. The key question is whether there is an acceptable, legitimate reason for the distinction. While these constitutional provisions do not frequently arise in a coastal context, in specialized circumstances they can be quite important.

REFERENCES

1. **Angell, J. K.,** *A Treatise on the Right of Property in Tidewaters,* 2nd ed., Charles C. Little and James Brown, Boston, 1847.
2. **Moore, S. A.,** *A History of the Foreshore and the Law Relating Thereto, and Hall's Essay on the Seashores of the Realm,* 3rd ed., Stevens and Haynes, London, 1888.
3. **Tillinghast, W. R.,** Tide-flowed lands and riparian rights in the United States, *Harvard Law Rev.,* 18, 341, 1905.
4. **Parsons, G. G.,** Public and private rights in the foreshore, *Columbia Law Rev.,* 22, 706, 1922.
5. **Taylor, W.,** The seashore and the people, *Cornell Law Q.,* 10, 303, 1925.
6. **David, H. P.,** The New York Law of the foreshore at the beginning of the 18th century, *Cornell Law Q.,* 11, 209, 1926.
7. **Littman, A. N.,** Tidelands: trusts, easements, custom and implied dedication, *Nat. Res. Law.,* 10, 279, 1977.
8. **Maloney, F. E. and Ausness, R. C.,** The use and legal significance of the mean high water line in coastal boundary mapping, *N.C. Law, Rev.,* 53, 185, 1974.
9. **Corker, C. E.,** Where does the beach begin and to what extent is this a federal question?, *Wash Law. Rev.,* 42, 33, 1966.
10. **Porro, A. A., Jr.,** Invisible boundary — private and sovereign marshland interests, *Nat. Res. Law.,* 3, 512, 1970.
11. **Nunez, P. K.,** Fluctuating shorelines and tidal boundaries; an unresolved problem, *San Diego Law Rev.,* 6, 447, 1969.
12. **Gay, N.,** High water mark: boundary between public and private lands, *U. Fla. Law Rev.,* 18, 553, 1966.
13. **Roberts, K.,** The Luttes case: locating the boundary of the seashore, *Baylor Law Rev.,* 12, 141, 1960.
14. In re Ashford, 50 Ha. 314, 440 P.2d. 76 (1968).
15. County of Hawaii v. Sotomura, 517 P.2d 57 (1973).
16. **Rice, D. A.,** Estuarine land of North Carolina: legal aspects of ownership, use and control, *N.C. Law Rev.,* 46, 779, 1968.

17. **Earnhardt, T. W.,** Defining navigable waters and the application of the public-trust doctrine in North Carolina: a history and analysis, *N.C. Law Rev.*, 49, 888, 1971.
18. **Schoenbaum, T. J.,** Public rights and coastal management, *N.C. Law Rev.*, 51, 1, 1972.
19. **Hill, D. P.,** Coastal wetlands in New England, *Boston U. Law Rev.*, 52, 724, 1972.
20. **Salsbury, S. M.,** Maryland's wetlands: the legal quagmire, *Md. Law Rev.*, 30, 240, 1970.
21. **Phillips, S. L.,** Regulation and ownership of the marshlands: the Georgia marshlands act, *Ga. Law Rev.*, 5, 563, 1971.
22. **Dawson, A. D.,** Protecting Massachusetts wetlands, *Suffolk U. Law Rev.*, 12, 755, 1978.
23. **McGregor, G. I., and Dawson, A. D.,** Wetlands and floodplain protection, *Mass. L. Rev.*, 64, 73, 1979.
24. **Jaffee, L. R.,** State citizen rights respecting greatwater resource allocation: from Rome to New Jersey, *Rutgers Law Rev.*, 25, 571, 1971.
25. **Anon.,** The public trust in tidal areas: a sometimes submerged traditional doctrine, *Yale Law J.*, 79, 762, 1970.
26. **Sax, J. L.,** Public trust doctrine in natural resource law: effective judicial intervention, *Mich. Law Rev.*, 68, 473, 1970.
27. Martin vs. Waddell, 41 U.S. 367 (1842).
28. Pollard's Lessee vs. Hagan, 44 U.S. 212 (1845).
29. Shively vs. Bowlby, 152 U.S. 1 (1894).
30. Illinois Central Railroad Co. vs. Illinois, 146 U.S. 387 (1892).
31. Marks vs. Whitney, 98 Cal. Report, 790, 491 P. 2d 374 (1971).
32. Hughes vs. Washington, 389 U.S. 290 (1967).
33. Trustees of Internal Improvement Fund vs. Ocean Hotels, Inc., 40 Fla. Supp. 26 (1974).
34. **Owens, D. W. and Brower, D. J.,** Public Use of Coastal Beaches, Report UNC-SG-76-08, University of North Carolina Sea Grant College Program, Raleigh, N.C., 1976.
35. **Owens, D. W.,** Public Rights In Shoreline Recreation Areas: A Selectively Annotated Bibliography, Report No. 894, Council of Planning Librarians, Monticello, Ill., 1975.
36. **Owens, D. W.,** The Law of Ownership and Use of Coastal Beaches, in *Access to the Nation's Beaches: Legal and Planning Perspectives*, Brower, D. J., Ed., Report UNC-SG-77-18, University of North Carolina Sea Grant Program, Raleigh, N.C., 1978, 19.
37. State ex rel. Thornton vs. Hay, 254 Ore. 584, 462 P.2d 671 (1969).
38. **McKeon, S. A.,** Public access to beaches, *Stanford Law Rev.*, 22, 564, 1970.
39. **Delo, L. E.,** The english doctrine of custom in Oregon property law, *Envir. Law*, 4, 383, 1974.
40. Gion vs. City of Santa Cruz, Dietz vs. King, 2 Cal. 3d 29, 84 Cal.Report 162, 465 P.2d 50 (1970).
41. **Berger, M. M.,** Nice guys finish last — at least they lose their property, *Cal. Western Law Rev.*, 8, 75, 1971.
42. **Gallager, J. V., Jure, C. W., and Agnew, G. E., Jr.,** Implied dedication: the imaginary waves of Gion-Dietz, *Southwestern U. Law Rev.*, 5, 48, 1973.
43. **O'Flaherty, M. A.,** This land is my land: the doctrine of implied dedication and its application to California beaches, *So. Cal. Law Rev.*, 44, 1092, 1971.
44. **Manzoni, C. R., Jr.,** A threat to the ownership of California beaches, *Santa Clara Law*, 11, 327, 1971.
45. **Burke, R. T.,** Public or private ownership of beaches: an alternative to implied dedication, *U.C.L.A. Law Rev.*, 18, 795, 1971.
46. **Finlay, S. P. and Van Til, D. J.,** Californians need beaches — maybe yours!, *San Diego Law Rev.*, 7, 605, 1970.
47. **Hayter, W. G.,** Implied dedication in California: a need for legislative reform, *Cal. Western Law Rev.*, 7, 259, 1970.
48. **Llewellyn, R. E. II,** The common law doctrine of implied dedication and its effect on the California coastline property owners, *Loyola (Los Angeles) Law Rev.*, 4, 438, 1971.
49. **Anon.,** Public access to beaches: common law doctrines and constitutional challenges, *N. Y. U. Law Rev.*, 48, 369, 1973.
50. Seaway Co. vs. Attorney General, 375 S.W.2d 923(Tex. Civ. App. 1964).
51. City of Daytona Beach vs. Tona-Rama, Inc., 294 So.2d 73 (Fla. 1974).
52. **Degnan, D. A.,** Public rights in ocean beaches: a theory of prescription, *Syracuse Law Rev.*, 24, 935, 1973.
53. **Agnello, T. J., Jr.,** Nonresident restrictions in municipally owned beaches: approaches to the problem, *Colombia J. Law Soc. Prob.*, 10, 177, 1974.
54. **Goldman, R. S.,** Access to public municipal beaches: the formulation of a comprehensive legal approach, *Suffolk U. Law Rev.*, 7, 936, 1973.
55. Borough of Neptune City vs. Borough of Avon-By-The-Sea, 61 N.J. 296, 294 A.2d 47 (1972).
56. **Anon.,** Water law — public trust doctrine bars discriminatory fees to nonresidents for use of municipal beaches, *Rutgers Law Rev.*, 26, 179, 1972.

57. Gewirtz vs. City of Long Beach, 69 Misc.2d 763, 330 N.Y.S.2d 495 (Sup. Ct. 1972), Aff'd. Mem., 358 N.Y.S.2d 957 (App. Div. 1974).
58. **Bryan, M. L.**, Which way to the beach? Public access to the beaches for recreational use, *S.C. Law Rev.*, 29, 627, 1978.
59. **Lafarge, J. F.**, Practical remedies to the public beach shortage, *Envir. Aff.*, 5, 447, 1976.
60. **Maloney, F. E., Fernandez, D., Parrish, A. R., Jr., and Reinders, J. M.**, Public beach access: a guaranteed place to spread your towel, *U. Fla. Law Rev.*, 29, 853, 1977.
61. **Ratliff, M.**, Public access to receding beaches, *Houston Law Rev.*, 13, 984, 1976.
62. **Zimmerman, R. H.**, Public beaches: a reevaluation, *San Diego Law Rev.*, 15, 1241, 1978.
63. In re Opinion of the Justices, 365 Mass. 681, 313 N.E.2d 561 (1974).
64. **Fairbank, W. A.**, Who owns the beach? Massachusetts refuses to join the trend of increasing public access, *Urban Law Ann.*, 11, 283, 1976.
65. DNR vs. Ocean City, 274 Md. 1, 332 A.2d 630 (1975).
66. **Nolan, J. J., Jr.**, Property — coastal shores, *U. Balt. Law Rev.*, 5, 349, 1976.
67. **Clark, J. R., Banta, J. S., and Zinn, J. A.**, Coastal Environmental Management: Guidelines for conservation of Resources and Protection Against Storm Hazards, U.S. Government Printing Office, Washington, D.C., 1980.
68. **Hildreth, R. G.**, Coastal natural hazards management, *Oreg. Law. Rev.*, 59, 201, 1980.
69. **Dolan, R., Hayden, B. P., Rea, C. C., and Heywood, J.** Shoreline erosion rates along the middle atlantic coast of the United States, *Geology*, 7, 602, 1979.
70. **Kaufman, W. and Pilkey, O. H.**, *The Beaches Are Moving*, Anchor Press/Doubleday, Garden City, N.Y., 1979.
71. **Schramm, W. E., Penland, S., Gerdes, R. G., and Nummedal, D.**, Effects of hurricane Frederick on Dauphin Island, Alabama, *Shore and Beach*, 48, 20, 1980.
72. **Horton, T.**, Relief from disaster relief: report from Dauphin Island, *Amicus J.*, 22, 1981.
73. Zabel vs. Tabb, 430 F.2d 199 (5th Cir. 1970).
74. **Barker, N. J.**, Sections 9 and 10 of the rivers and harbors act of 1899: potent tools for environmental protection, *Ecology Law Q.*, 6, 109, 1976.
75. **Martin, K. R.**, Wetlands controversy: a coastal concern washes inland, *Notre Dame Law*, 52, 1015, 1977.
76. Kaiser Aetna vs. United States, 444 U.S. 164 (1979).
77. **McGuinness, M.**, Navigable water not always subject to free public access, *Natural Res. J.*, 21, 161, 1981.
78. **Hanks, E. H. and Hanks, J. L.**, An environmental bill of rights: the citizen suit and the national environmental policy act of 1969, *Rutgers Law Rev.*, 24, 230, 1970.
79. **Leventhal, H.**, Environmental decisionmaking and the role of the courts, *U. Pennsylvania Law Rev.*, 122, 509, 1974.
80. **McGarity, T. O.**, The courts, the agencies, and NEPA threshold issues, *Texas Law Rev.*, 55, 801, 1977.
81. **Jacobson, K. A.**, Environmental law — judicial review under NEPA, *Villanova Law Rev.*, 23, 1170, 1978.
82. **Munns, M. J.**, National environmental policy act: an ambitious purpose; A Partial Demise, *Tulsa Law J.*, 15, 153, 1980.
83. **Liroff R. A.**, NEPA litigation in the 1970s: a deluge or a dribble, *Natural Res. J.*, 21, 315, 1981.
84. Calvert Cliffs' Coordinating Committee vs. Atomic Energy Commission, 449 F.2d 1109 (D.C. Cir. 1971).
85. **Kuehn, R. R.**, The shifting sands of federal barrier islands policy, *Harvard Environ. Law Rev.*, 5, 217, 1981.
86. **Douglas, P. M. and Petrillo, J. E.**, California's coast: the struggle today — a plan for tomorrow, *Fla. St. U. Law Rev.*, 4, 177, 1976.
87. **Finnell, G. L., Jr.**, Coastal land management in California, *Am. Bar. F. Res. J.*, 1978, 647, 1978.
88. **Davidson, J. M.**, Coastal zone management and planning in California: strategies for balancing conservation and development, *Urban Law Annual*, 15, 253, 1978.
89. **Finnell, G. L., Jr.**, Saving paradise: the Florida environmental land and water management act of 1972, *Urban Law Ann.*, 6, 103, 1973.
90. **Finnell, G. L. Jr.**, Coastal land management in Florida, *Am. Bar F. Res. J.*, 1980, 303, 1980.
91. **Pelham, T. G.**, Regulating areas of critical state concern: Florida and the model code, *Urban Law Ann.*, 18, 3, 1980.
92. **Schoenbaum, T. J.** The management of land and water use in the coastal zone: a new law is enacted in North Carolina, *N.C. Law Rev.*, 53, 275, 1974.
93. **Glenn, P. G.**, The coastal area management act in the courts: a preliminary analysis, *N.C. Law Rev.*, 53, 303, 1974.
94. **Heath, M. S., Jr.**, A legislative history of the coastal area management act, *N.C. Law Rev.*, 53, 345, 1974.
95. **Owens, D. W.**, Public participation in local land use planning: concepts, mechanisms, state guidelines and the coastal area management act, *N.C. Law Rev.*, 53, 975, 1975.

96. **Schoenbaum, T. J. and Rosenberg, R. H.**, The legal implementation of coastal zone management: the North Carolina model, *Duke Law J.*, 1976, 1, 1976.
97. **Schoenbaum, T. J. and Silliman, K. G.**, Coastal planning: the designation and management of areas of critical environmental concern, *Urban Law Ann.*, 13, 15, 1977.
98. **Hershman, M. J., and Mistric, M. M.**, Coastal zone management and state-local relations under the Louisiana Constitution of 1974, *Loyola Law Rev.*, 22, 273, 1975.
99. **Forman, W. H., Jr.**, Louisiana coastal resources management act of 1978, *La. Bar J.*, 28, 91, 1980.
100. **Marcel, K. W. and Bockrath, J. T.**, Regional governments and coastal zone management in Louisiana, *La. Law Rev.*, 40, 887, 1980.
101. **Goodman, J. M.**, The Delaware coastal zone experience, *Environ. Law*, 5, 727, 1975.
102. **Anon.**, Assimilating human activity into the shoreline environmental: the Michigan shoreland protection and management act of 1970, *Iowa Law Rev.*, 62, 149, 1976.
103. **Wyche, B. W.**, South Carolina coastal zone management act of 1977, *S.C. Law Rev.*, 29, 666, 1978.
104. CEED vs. Coastal Commission, 43 Cal. App.3d 306, 118 Cal.Rptr. 315 (Ct. App. 1974).
105. Toms River Affiliates vs. Dept. of Environmental Protection, 140 N. J. Super. 135, 355 A.2d 679, Cert. Denied, 71 N. J. 345, 364 A.2d 1077 (1976).
106. Adams vs. Dept. of Natural and Economic Resources, 295 N.C. 683, 249 S.E.2d 402 (1978).
107. Askew vs. Cross Keys Waterways, 372 So.2d 913 (Fla., 1978).
108. National Atmospheric and Oceanic Administration, Office of Coastal Zone Management, The First Five Years Of Coastal Zone Management, U.S. Department of Commerce, Washington, D.C., 1979.
109. **Burby, R. J. and French, S. P.**, Coping with floods: the land use management paradox, *Am. Plan. Assoc. J.*, 47, 289, 1981.
110. **Maloney, F. E. and O'Donnell, A. J., Jr.**, Drawing the line at the oceanfront: the role of construction setback lines in regulating development of the coastal zone, *U. Fla. Law Rev.*, 30, 383, 1978.
111. **Shows, E. W.**, Florida's coastal setback line — an effort to regulate oceanfront development, *Coastal Zone Manage. J.*, 4, 151, 1978.
112. Kopetzke vs. County of San Mateo, 396 F. Supp. 1004 (1976).
113. Beri vs. Salishan Properties, Inc., 282 Or. 569, 580 P.2d 173 (1978).
114. **Hagman, D. G. and Misczynski, D. J., Eds.**, *Windfalls for Wipeouts: Land Value Capture and Compensation*, Am. Soc. of Plan. Officials, Chicago, 1978.
115. **Godschalk, D. R., Brower D. J., McBennett, L. D., and Vestal, B. A.**, *Constitutional Issues of Growth Management*, Am. Soc. of Plan. Officials, Chicago, 1977.
116. Pumpelly vs. Green Bay Co., 80 U.S. 166 (1871).
117. Lorio vs. Sea Isle City, 88 N.J. Super. 506, 212 A.2d 802 (1965).
118. Mugler vs. Kansas, 123 U.S. 623 (1887).
119. Pennsylvania Coal Co. vs. Mahon, 260 U.S. 393 (1922).
120. Miller vs. Schone, 276 U.S. 272 (1928).
121. Nectow vs. City of Cambridge, 277 U.S. 183 (1928).
122. **Dunham, A.**, Flood control via the police power, *U. Penn. Law Rev.*, 107, 1098, 1959.
123. **Sax, J. L.**, Takings and the police power, *Yale Law J.*, 74, 36, 1964.
124. **Michelman, F. I.**, Property, utility and fairness: comments on the ethical foundations of "Just Compensation" law, *Harvard Law Rev.*, 80, 1165, 1967.
125. **Van Alstyne, A.**, Taking or damaging by police power: the search for inverse condemnation critera, *So. Cal. Law Rev.*, 44, 1, 1970.
126. **Sax, J. L., Takings,** Private property and public rights, *Yale Law J.*, 81, 149, 1971.
127. **Kingham, R. F.**, State and local wetlands regulation: the problem of taking without just compensation, *Va. Law Rev.*, 58, 876, 1972.
128. **Bosselman, F., Callies, D., and Banta, J.**, *The Taking Issue*, U.S. Government Printing Office, Washington, D.C., 1973.
129. **Plater, Z. J. B.**, The takings issue in the natural setting: floodlines and the police power, *Texas Law Rev.*, 52, 201, 1974.
130. Gorieb vs. Fox, 274 U.S. 603 (1927).
131. King vs. Inc. Village of Ocean Beach, 207 Misc. 100, 136 N.Y.S.2d 609, (Sup. Ct. 1954), Aff'd 286 App. Div. 850, 143, N.Y.S.2d 637 (1955).
132. Lemp vs. Town Bd. of Islip, 90 Misc.2d 360, 394 N.Y.S.2d 517 (Sup. Ct. 1977).
133. Speilgle vs. Borough of Beach Haven, 46 N.J. 479, 218 A.2d 129 (1966), Cert. Denied, 385 U.S. 831 (1966), 166 N.J. Super. 148, 281 A.2d 377 (App. Div. 1971).
134. Inhabitants of Town of Boothbay Harbor vs. Russell, 410 A.2d 554 (Me., 1980).
135. Town of Indialantic vs. McNulty, 400 So.2d 1227 (Fla. App. 1981).
136. **Rice, D. A.**, "Taking" by Regulation, and the North Carolina Coastal Area Management Act, Report UNC-SG-75-26, University of North Carolina Sea Grant College Program, Raleigh, N.C., 1976.
137. Just vs. Marinette Co., 56 Wis.2d 7, 201 N.W.2d 761 (1972).
138. Graham vs. Estuary Properties, Inc., 399 So.2d 1374 (Fla., 1981).

Chapter 9

THE GREAT LAKES: A MICROCOSM OF THE WORLD OCEAN

Robert A. Ragotzkie

TABLE OF CONTENTS

I. INTRODUCTION

The freshwater seas that are called the Laurentian Great Lakes have both lacustrine and oceanic characteristics. As bounded bodies of freshwater which are hydrologically controlled by their watersheds and drainage system, these seas are clearly lakes. However owing to their size they display many oceanic characteristics. Their resources are as diverse as the salty seas and human use of these resources is as intense or more so than along the most densely populated seacoasts.[1] They are also international; four of the five Great Lakes are shared by the U.S. and Canada. The study of these inland seas and their resources requires techniques and approaches analogous to those used for the world ocean. Because the Great Lakes and their resources are shared by eight states, namely New York, Pennsylvania, Michigan, Ohio, Indiana, Illinois, Wisconsin, and Minnesota, the Province of Ontario, the U.S. and Canada, and several native American tribes, their management is complex. Over the years a variety of international, interstate, intrastate, and state-local institutional arrangements have been developed. In many respects the approaches to the management of the Great Lakes have evolved faster and have achieved a more advance stage of development than for the oceans.

II. HYDROGRAPHIC CHARACTER OF THE GREAT LAKES

A. Basins

By oceanic standards the Great Lakes (Figure 1) are small, having a total area of 244,000 km^2 or about 1/4 of 1% of the world ocean. They are also shallow, their maximum depth ranging from 64 m for Lake Erie to 406 m for Lake Superior (Table 1). By lake standards they are large; one-fifth of the surface freshwater of the earth is contained by these lakes. However the relative dimensions of the Great Lakes and the oceans are remarkably similar. For example comparison of the largest and deepest ocean, the Pacific, with the largest and deepest Great Lake, Superior, reveals that both are about 4,000 times as wide as their mean depths and 1,500 times as wide as their maximum depths. Though these ratios are most certainly fortuitous, there being no similarity in the geological processes responsible for the formation of these two basins, the morphological similarity does place the dimensions of the Great Lakes in perspective.

Unlike the ocean basins which are primarily tectonic in origin with depositional processes dominating coastal regions, the Great Lakes basins, with the exception of the Lake Superior basin which is primarily tectonic in origin, were formed primarily by glacial processes. The Great Lakes are geologically very young, approaching their present form barely 20,000 years ago while the oceans' age is measured in hundreds of millions of years.

B. Circulation

As in the oceans the circulation of the Great Lakes is primarily wind-driven. Currents in the Great Lakes tend to be parallel to the coasts. Ekman transport and the resulting coastal upwelling or downwelling leads to strong horizontal temperature gradients normal to the coast and within a half pendulum day or so the coastal current becomes geostrophic. These coastal currents usually only persist for a few days to a week, but sometimes last for several weeks. The overall circulation patterns take the form of gyres, either encompassing an entire lake as frequently occurs in Lake Ontario or as several gyres in Lakes Michigan and Superior.

In Lake Superior an eastward flowing current along the Keweenaw Peninsula, on the south side of the lake, persists for at least six months from about June through November and may continue year round. This current, called the Keweenaw Current,[2] is quite steady, extends from the surface to 40 to 60 m deep. It can extend up to 20 km from the coast but is more frequently about 10 km in width. Its velocity ranges from 25 to 90 cm sec^{-1} or nearly two knots.

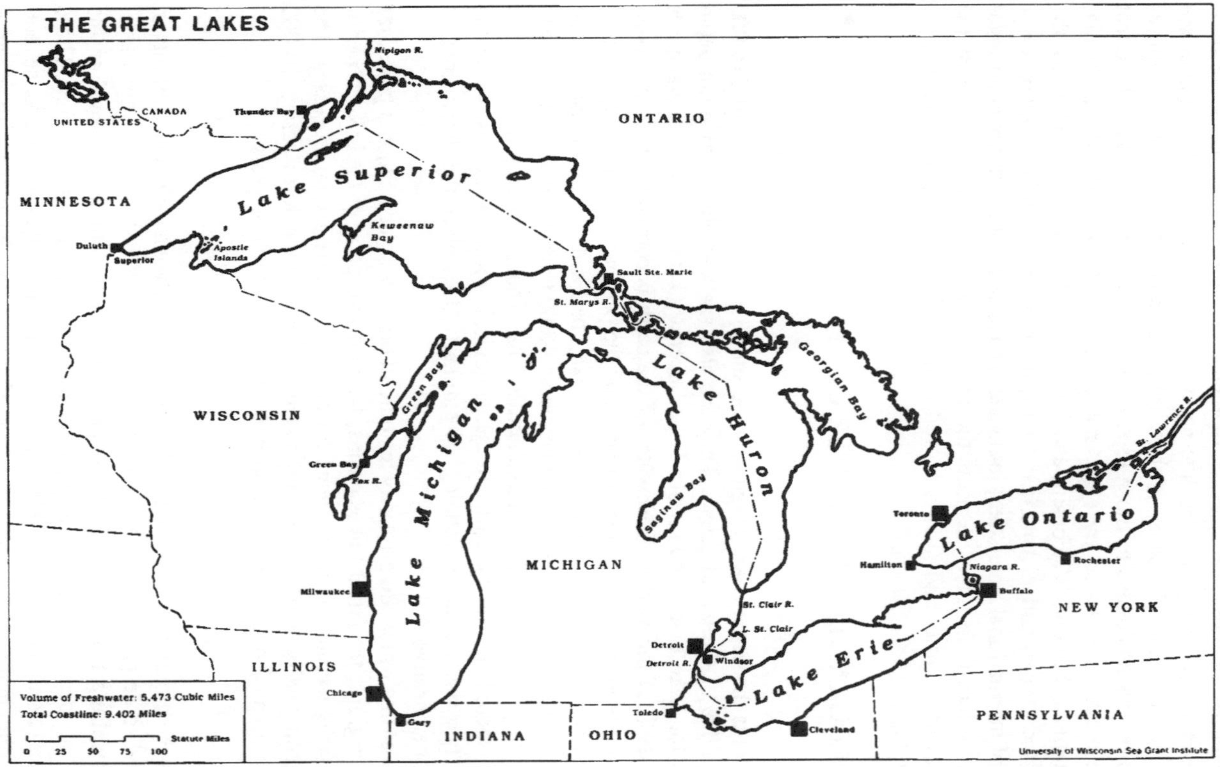

FIGURE 1. The Laurentian Great Lakes showing key ports and the international boundary.

Table 1
PHYSICAL CHARACTERISTICS OF THE GREAT LAKES

Lake	Area lake (km^2)	Area drainage basin land (km^2)	Mean depth (m)	Maximum depth (m)	Elevation above sea level (m)	Mean discharge (m^3/sec)
Superior	82,103	125,356	149	406.3	183.6	2,076
Michigan	57,757	118,104	85	281.3	176.9	1,558
Huron	59,570	131,313	59.4	228.6	176.9	5,038
Erie	25,667	58,793	18.9	64	174.4	5,545
Ontario	19,010	63,973	86.3	244.4	75	6,624

Extensive studies have been made of this current by means of moored current meters, drogues, temperature cross-sections, aerial photography of drift cards, infrared radiometry, and infrared imagery.[2-7] During the summer the current is strongly baroclinic with a cross current thermal gradient of 12°C or more. The dynamics of the Keweenaw Current are surprisingly similar to those of the Gulf Stream. Both are essentially thermal boundary currents. Cross stream velocity shears are up to 10 times greater in the Keweenaw Current than in the Gulf Stream. Eddies of 100 to 500 m in diameter occur on the outer edge of the Keweenaw but unlike the Gulf Stream eddies are relatively short-lived, lasting 30 min or so. Transfer of kinetic energy both to and from these eddies to the main flow occur in both systems.

In contrast to the Gulf Stream the Keweenaw Current is seasonal in nature, exhibiting a fairly regular development from a spring coastal current, to a baroclinic current in summer, and finally to a wider, somewhat slower, and less baroclinic current later in the season. The winter situation is largely unknown owing to the difficulty of conducting field observations while the shores and harbors are icebound, even though the region of the current itself is largely ice-free. Visual observations of ice movement from west to east during the winter suggest that this current may be a year-round feature of Lake Superior's circulation.[8]

Also the Keweenaw Current being a relatively small east-west system does not include the planetary aspects of the Gulf Stream, such as the latitudinal variation of the Coriolis term. Nevertheless the similarities are interesting and many of the same observational techniques have been used to study both systems.[4,5] Interestingly even the first mentions of both currents are similar, the Gulf Stream being reported by early mariners and later in more detail by New England fishermen,[9] and the Lake Superior Keweenaw Current first mentioned by Great Lakes fishermen.[10]

Though the great thermohaline circulation of the ocean has no analogy in the Great Lakes because of their lack of salt and relative shallowness, gravity-driven density currents occur regularly in Lake Superior and probably also in Lake Huron. These vertical currents are caused by the flow of cold, sediment-laden water from streams carrying large volumes of spring meltwater. The combination of low temperature and high sediment load causes water densities greater than that of the cold lake water. These currents may descend 100 m or more along the steeply sloping bottom of the north shore of Lake Superior, effectively transporting sediments and stream water to the deeper regions of the lake.

In Lake Superior a region of very cold (4 to 6°C) surface water persists during much of the summer despite intense heating by solar radiation. An analysis of the heat balance of the region based on the calculated solar heating rate and the observed water temperatures shows that upwelling in the "cold pool" averages about 4 m/day during the heating season.[11] This vertical motion is apparently maintained by a large cyclonic gyre in the central portion of the lake. The Keweenaw Current which is on the south side of this gyre probably plays a major role in its maintenance.

Coastal upwelling is also a familiar phenomenon in the Great Lakes sometimes reaching 200 m in vertical extent along the north shore of Lake Superior.[12,13] Upwelling frequently occurs along the Keweenaw coast[6,7] and along Wisconsin's Lake Michigan coast. These vertical motions are usually associated with Ekman transport occurring when strong winds blow parallel to the coast.

C. Waves and Tides

The tides of the ocean have their analogy in the wind-driven seiches of the Great Lakes. These surface oscillations which are always present may attain vertical ranges of 2 m or more during persistent wind episodes and must often be taken into account by ship captains when entering harbors with entrance channels of limited depth. Mortimer who has studied these oscillations in detail[14,15] has demonstrated that these seiches rotate much like tidal waves. However they are more complex in that several (up to eight) modes of oscillation can occur simultaneously. Power spectrum analysis of these surface oscillations reveals not only the presence of multiple free modes of oscillation but also the presence of diurnal and semidiurnal tides (Figure 2).

Lunar tides also occur in the Great Lakes. Father André, a Jesuit missionary who explored Green Bay described in detail "the tides of the Bay de Puants [Green Bay]."[16] Actually the water level fluctuations he described were mostly wind-driven seiches which, in Green Bay, have a natural period close to the semidiurnal tides of the ocean with which he was familiar. Later analyses by Mortimer[14] however showed that although the seiche was the dominant cause of the surface oscillations, lunar spring tides of 0.1 to 0.2 ft occur in Lake Michigan and up to 0.6 ft in Green Bay.

The semidiurnal tides rotate within the lake basins with amphidromic patterns much like that of the basins of the oceans (Figure 3). Rotation is clockwise in Lake Superior but is counterclockwise in Lake Michigan. The seiche wave, on the other hand, rotates in a counterclockwise direction in both basins.[14]

Wind waves of the Great Lakes do not of course attain the huge dimensions of those in a North Pacific cyclone, but fall storms frequently generate waves in excess of 16 ft in Lake Superior and 40 footers have been reported. Great Lakes storms are particularly fearsome, however, both because of the suddenness with which they can develop and the steepness of the waves themselves. Coupled with subfreezing temperatures which cause instant freezing of freshwater spray on hulls and superstructure, the short steep waves and hurricane-like winds of an early winter cyclonic storm have sunk literally thousands of Great Lakes vessels, from early schooners and steamboats to modern 700-ft ore carriers like the Edmund Fitzgerald which was lost in Lake Superior during a November storm of 1977.

III. GREAT LAKES RESOURCES

A. Overview

The Great Lakes are a unique and valuable resource, lying within the most heavily populated and industrialized region of the nation. As such they are under intense pressure to provide clean drinking water for nearly 25 million U.S. citizens, high quality industrial water for a wide variety of basic product manufacturing plants, cooling water for electric power plants, and recreational opportunities for over 60 million people. They provide an efficient and cheap water transportation system for iron, coal, limestone, and grain. They support an immensely productive fishery which is exploited by both commercial and sport fishing, and they are a sink for domestic and industrial wastes. Taken all together, these resources constitute the backbone of the industrial and population heartland of the U.S. and Canada.

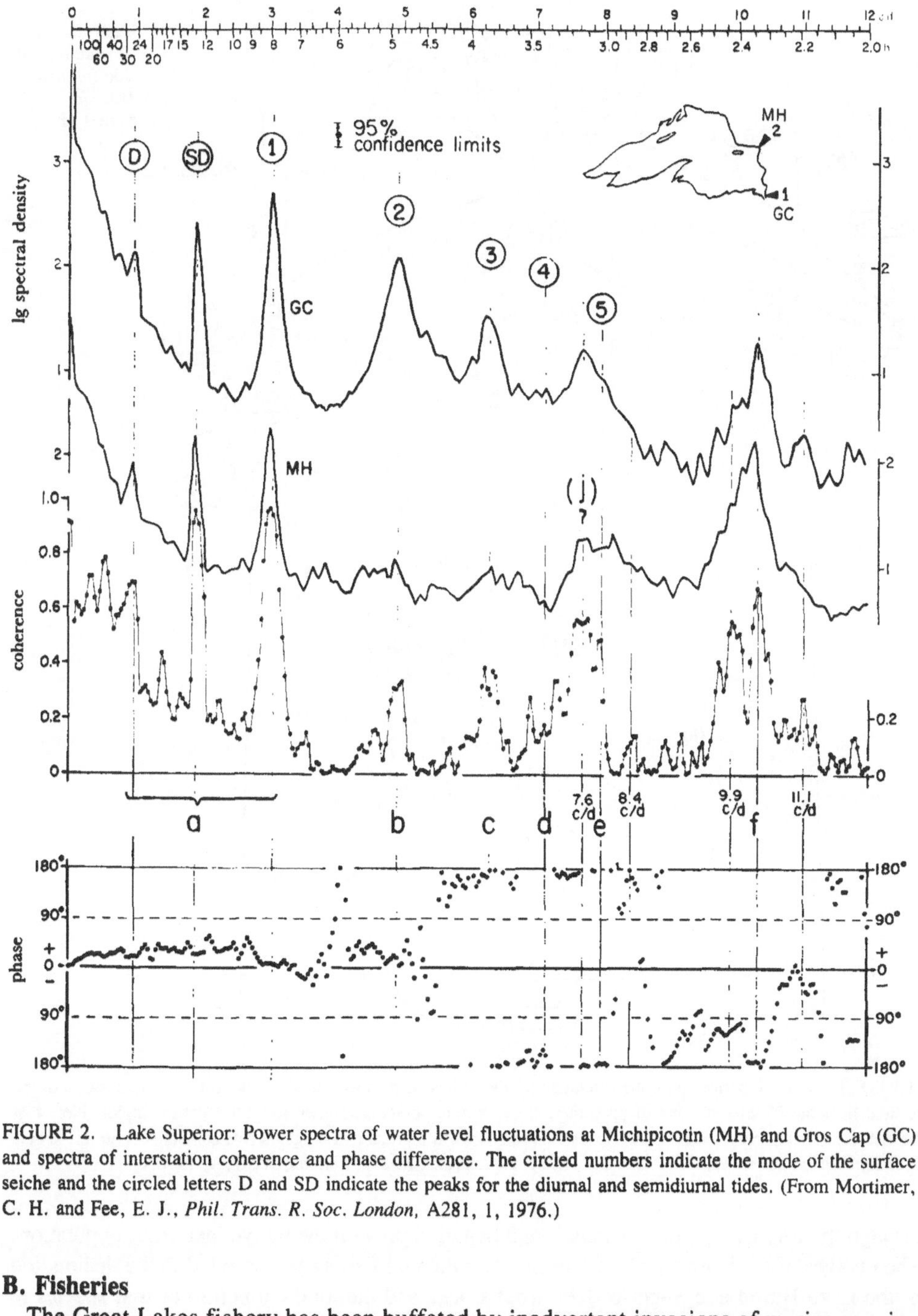

FIGURE 2. Lake Superior: Power spectra of water level fluctuations at Michipicotin (MH) and Gros Cap (GC) and spectra of interstation coherence and phase difference. The circled numbers indicate the mode of the surface seiche and the circled letters D and SD indicate the peaks for the diurnal and semidiurnal tides. (From Mortimer, C. H. and Fee, E. J., *Phil. Trans. R. Soc. London*, A281, 1, 1976.)

B. Fisheries

The Great Lakes fishery has been buffeted by inadvertent invasions of marine species followed by large scale experimental introduction of several species of Pacific and Atlantic salmon plus several nonnative species of trout. As a result the fish population and assemblage of species of today is quite different from that of a century or so ago. Though most of the native fishes still exist, they no longer dominate the system, and a few have become extinct in some of the lakes.

The chain of events leading to the present situation began in the late 1930s when the sea lamprey, a marine parasitic fish, entered the upper Lakes, Huron, Michigan, and Superior,

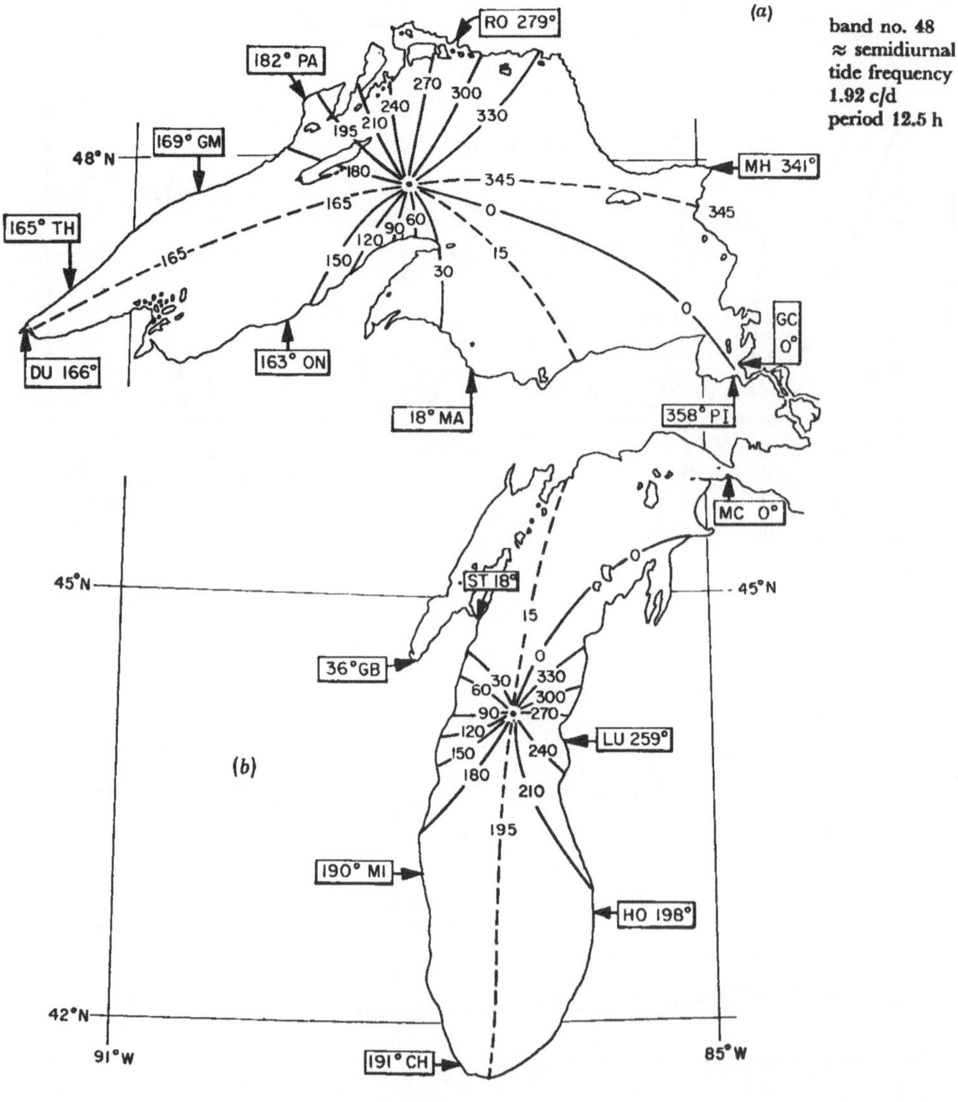

FIGURE 3. Lake Superior and Lake Michigan: Station phase angles for the semidiurnal tide. The position of the co-tidal lines at 30° intervals and of presumed amphidromic points are drawn to fit the station angles. Note that the tide in Lake Superior rotates in the clockwise direction while in Lake Michigan it rotates in the counterclockwise direction. (From Mortimer, C. H. and Fee, E. J., *Phil. Trans. R. Soc. London,* A281, 1, 1976.)

through the improved Welland Canal, and began to prey on the native lake trout population. The ravages of the lamprey plus increased commercial fishing pressure led to the decimation of the Lake Huron and Superior lake trout stocks and the total extinction of that species in Lake Michigan by the early 1950s.[17] With the disappearance of the lake trout, the major predator, the alewife, a marine forage fish, which had been in the St. Lawrence River and Lake Ontario for a long time, moved into the upper lakes and began to proliferate. Their numbers increased explosively and the annual die-offs of this herring-like fish created intolerable nuisance conditions along the shores.

As a management measure the state of Michigan began to stock coho salmon, a Pacific species, in Lake Michigan to control the alewife population. The experiment worked, and other states soon followed suit. Other species were added later. There is now a very large program throughout all the Great Lakes under which 25 million Pacific coho and chinook salmon, Atlantic salmon, several species of nonnative trout and native lake trout are stocked annually. Meanwhile lamprey control measures have taken effect,[18-20] and as a result the Great Lakes now support a highly productive salmon and trout fishery. However, as the tributary streams cannot provide a proper spawning and nursery ground for the salmon and because of intense exploitation of both salmon and trout, the fishery remains dependent on continued stocking for its productivity.

Unfortunately during this same period the lakes became contaminated with a number of man-made organic substances, some of which, because of their high fat solubility, tended to concentrate in fat-rich fish like alewives, salmon, and trout.[21] Notable of these were DDT, PCBs, and mirex. Even though this contamination was relatively low in concentration, the health risk to man was uncertain. As a result these fish were embargoed for commercial use. Thus the salmon and trout fishery has become primarily recreational in character, and use of the abundant alewife was limited to meal for animal foods.

The Great Lakes commercial fishery has declined somewhat but is now stabilizing. It is producing about $40 million/year dockside value or an estimated $160 million total value per year.[22] The fishery includes perch, walleye and rough fish from the lower lakes and whitefish, chubs, perch, herring, and smelt from the upper lakes. A limited commercial lake trout fishery exists in Lake Superior, and there is a potential for a larger and very valuable lake trout fishery in Lakes Michigan and Huron once the microcontaminants problem is solved.

In the past 15 years the Great Lakes sport fishery has blossomed into a major industry, with an estimated annual value to the region of $1 billion.[22] This explosive growth came about first by the introduction of salmon and the increased stocking of several species of trout first in the upper lakes and ultimately in all five lakes. More recently the revitalization of the walleye fishery in Lake Erie and also in Green Bay, Lake Michigan has added another major sportfishing resource. The phenomenal development of the Great Lakes sportfishing industry has stimulated tourism, increased boat and motor sales, and most importantly has sharply raised the public consciousness of the Great Lakes as a major regional resource. As a result public pressure has influenced government agencies to pay much more attention to water quality, coastal enhancement, as well as to the fisheries themselves. It has become increasingly clear that the waters of the Great Lakes have an exceedingly high biological productivity. The lakes have the capacity to maintain a high sustained yield of high quality protein. We are only beginning to learn how to manage and preserve this system to take advantage of this capacity.

C. Water

The freshwater of the Great Lakes is a unique resource, unparalleled anywhere in the world. Measuring nearly 5,500 mi^3 in volume the Great Lakes contains nearly one-fifth of the earth's surface fresh water. The water is of exceptionally high quality, and is used for drinking by about 25 million U.S. citizens. Though the process of eutrophication or over-enrichment by plant nutrients has somewhat degraded the water of Lake Erie and threatened Lakes Michigan and Ontario,[23] a massive effort to improve the treatment of domestic and industrial wastes has begun to take effect. The eutrophication process has been stopped in most areas and the trend may have been reversed in the lower lakes.[24,25] Nonpoint sources of nutrients, namely runoff from agricultural lands and unsewered urban areas remain a major problem and efforts to alleviate this source are proceeding more slowly.

In the past 15 years or so the problem of microcontaminants like heavy metals and man-made organic chemicals has become a major concern. In general the concentrations of most of these substances in the water are still too low to be a threat to the human health through drinking water. However many of these substances are taken up by aquatic organisms and some tend to concentrate as they move up the food chain. As a result, the fish, particularly the top predators like salmon and trout and bottom feeders like carp and catfish, have become contaminated to the point where they are potentially hazardous for regular human consumption.[21] This problem is most acute in Lake Michigan where the concentration of PCBs (polychlorinated hydrocarbons) in adult salmon and trout often exceed the USFDA tolerance level of 5 ppm. In Lake Ontario fish contamination by mirex, a pesticide, has been a serious problem. There are some indications that the PCB problem is abating, but atmospheric deposition of this and other undesirable substances remains a major source of contamination.[26]

Despite these problems water quality remains high, especially when compared to most large rivers and reservoirs. Industrial use of the water is extensive. Manufacturing plants as well as basic industries like steelmaking have been established on the coasts in order to take advantage of the cheap and abundant water supply. Because of their requirements for large volumes of cooling water many electric power generating plants, both nuclear and fossil fuel, have been sited on Great Lakes coasts.

The Great Lakes serve as a sink for the domestic and industrial wastes of communities on their shores and within the basin. When one considers that over 50 million Americans and Canadians live within the Great Lakes basin and that much of the basin is highly industrialized, the magnitude of the waste disposal problem becomes apparent. Though the lakes have an enormous capacity to assimilate ordinary organic wastes, this capacity is now in danger of being exceeded, particularly near the large urban centers like Chicago, Milwaukee, Detroit, Cleveland, Toronto, and Buffalo-Niagara Falls. The toxic contaminant problem is particularly acute because many of these substances, like PCBs and heavy metals are not biologically degraded into harmless end-products. Even so, given time and with the reduction of the input of these contaminants, the biological and physical processes within the lakes will transport these substances to the sediments where they will ultimately be buried out of harms way (Figure 4). Thus the lakes perform a valuable waste disposal function for a large, highly populated and industrialized region. The challenge to our stewardship is to use this waste disposal capacity in such a way that we do not degrade or destroy the other resources of the lakes such as fisheries, drinking water, and recreational use.

Still in the future is the possibility of diverting Great Lakes water outside the basin to meet water needs in arid regions of the great plains. Energy projects like coal gasification in the Dakotas and Wyoming and oil shale development have already begun to look toward the Great Lakes for water. Agriculture which now depends on the great Ogalalla aquifer underlying the high plains for irrigation water will eventually need another source of water when the groundwater runs out. Whether the Great Lakes can or should be tapped for this purpose is a question that will almost certainly have to be faced in the future. The economic, legal, institutional, and environmental dimensions of this question have hardly even begun to be examined.[27]

D. Other Resources

1. Minerals

At present the only minerals being extracted from the bottom of the lakes are sand, gravel, and gas. Salt is being mined from beneath Lake Erie through lateral tunnels from land based mines. There are known deposits of copper under Lake Superior but economic and environmental factors preclude production of the metal. There is also oil under Lake Erie and probably Lake Michigan. However the amounts are not thought to be great, and no major plans for recovering it are underway at this time.

Iron ore, limestone, and coal are probably the most important mineral resources of the Great Lakes basin. Even though these minerals are not being extracted from beneath the lakes themselves, the lakes have played a key role in their exploitation. The major iron

FIGURE 4. PCBs ultimately end up in the bottom sediments. A box-core sampler shown here is being lowered to sample the sediments of Lake Michigan for later laboratory analysis of PCBs. University of Wisconsin Sea Grant photo.

deposits of the Mesabi Range are located just north of Lake Superior, and limestone occurs all along the coasts of northern Lake Michigan and Lake Huron. Abundant coal deposits occur along the southern border of the lakes, extending across Illinois, Indiana, Ohio, and Pennsylvania. These three minerals, iron, limestone, and coal, form the basis of steelmaking. Their abundance in the Great Lakes basin and the cheap Great Lakes water transportation which links them have been the key to the development of the major steel industry of the region. This steel industry, the availability of cheap coal and water and direct access to a water transportation link to the Atlantic Ocean have in turn resulted in the creation of the major industrial complex extending along the entire length of the Great Lakes.

2. Energy

As noted earlier in this section extensive generation of electric power along the coasts of the Great Lakes is made possible by virtue of the availability of cooling water and cheap transportation of coal. With the increased exploitation of western coal the Great Lakes and St. Lawrence Seaway are providing a major water link to the Atlantic coast. In 1978, 37% of U.S. coal exports originated from Great Lakes ports. Lake Erie has been producing gas for several decades and new wells are now being drilled. Finally the lakes provide a substantial amount of hydropower at Sault Ste. Marie, Niagara Falls, and the St. Lawrence Seaway.

3. Recreation and Coastal Resources

As on the saltwater seacoasts recreation is the fastest growing Great Lakes industry and may now be the largest. Though an accurate estimate of its economic value has not been made, this industry is clearly in the multibillion dollar range. In Wisconsin alone, with access to two Great Lakes, the value of the recreation industry in the coastal counties generated gross sales of $2.5 billion in 1980. For a more complete discussion of marine recreation, see Chapter 1 by Rorholm.

The Great Lakes have a combined coastline in the U.S. and Canada of 9,400 mi. These coasts range from low-lying marshes, sand beaches and bluffs to rocky shores, island studded passages, and spectacular escarpments.

The uses of this extensive and varied coast are typical of the saltwater seacoasts, namely ports, industry, highly prized residential use, recreation, fisheries, wetlands, fish spawning and nursery areas, and extensive natural areas. The public policy and management approach for maintaining a balanced and healthy use of this coast is discussed more completely by Miller in Chapter 7 and Owens in Chapter 8.

E. Resource Use Pressure and Need for Management

From this brief overview it is clear that the Great Lakes are subject to the same array of use pressures as the oceans, particularly the coastal regions. As closed bodies of water with relatively shallow basins the Great Lakes have a much smaller capacity to absorb and resist these pressures. Thus conflicts of resource allocation and environmental degradation have occurred much sooner and escalated more rapidly than in the oceans; the impact of man has been accelerated. However the in-lake processes are the same, and it is reasonable to postulate that the human responses and institutional processes are also similar.

As ecosystem changes occurred, water quality degraded and fisheries collapsed, the need for pollution abatement, fishery management, and the control of coastal development was clear and immediate. The societal responses have been sharp and often controversial, but overall the people of the Great Lakes region have begun to recognize the need for management of the Great Lakes as a whole.

IV. MANAGEMENT OF THE GREAT LAKES

A. International

Compared to the world ocean, the problem of establishing a management framework for the Great Lakes is relatively straightforward. Unlike in the oceans where the complexity of bringing together 150 nations in the Law of the Sea Conference in an attempt to reach agreement on how to preserve the integrity of the marine environment and manage ocean resources, the Great Lakes are shared by two friendly nations with a common interest in safeguarding the quality of their boundary waters. Even considering the apparent simplicity of the Great Lakes situation and the common interest of the parties, the degree of agreement and level of management sophistication that has been achieved is remarkable.

There is no doubt that the common cultural heritage of the U.S. and Canada has been a major factor in the achievement of a common purpose in their approach to the Great Lakes and their resources. Beyond the favorable political and cultural climate, however, the more rapid response of the lakes to human impact has forced the governments of both nations to move sooner and more quickly to create institutions and methods to deal with the situation.

1. International Joint Commission

The basis of international management of the Great Lakes is the Boundary Waters Treaty of 1909 between the U.S. and Canada.[28] This treaty established the International Joint Commission to deal with the international problems of the Great Lakes and other boundary waters. The Commission, composed of three Canadians and three Americans, conducts its business as a single body not as separate national delegations. The Commission has three primary functions:

1. Regulatory — it approves or disapproves applications from government, companies or individuals for obstructions, uses or diversions of water which affects the natural level or flow of water.
2. Investigative — it investigates matters of difference along the common frontier, reports the facts and recommends appropriate action by the two governments.
3. Surveillance and Coordination — It monitors compliance with the terms and conditions of orders of approval, including the objectives and standards set forth in the Water Quality Agreements of 1972 and 1978.

A fourth function which has never been utilized permits governments to refer any issues to the Commission for binding decision.

Over the years a number of boards have been set up to advise the Commission. Among these are the Great Lakes Levels Board, the Water Quality Board, and the Science Advisory Board. The latter two are primarily concerned with the Water Quality Agreements of 1972 and 1978.[29] These agreements have set both general and specific goals for the maintenance of water quality of the lakes. The stated purpose of the 1978 Agreement is "to restore and maintain the chemical, physical, and biological integrity of the waters of the Great Lakes Basin Ecosystem."

The key phrase in this broad statement is "Great Lakes Basin Ecosystem." By including this term the two nations, the U.S. and Canada, have agreed to manage the entire basin: the land surrounding the lakes, the streams flowing into them, and the lakes themselves. The Agreement goes far beyond water quality alone; it calls for the restoration of the integrity of the entire ecosystem: water, land, air, and living things, including man.[30]

Of course, a statement of purpose, however grand, has little real effect on the lake system. Implementation is the major challenge. Translation of the goal of restoring and maintaining the ecosystem to real policies, environmental standards, and finally management action is proving to be more difficult than was anticipated. It is a relatively straightforward task to monitor a water body with regard to specific substances, say heavy metals or nutrients, and to devise effluent controls to meet sepcified standards. To define the effects of these substances on the ecosystem as a whole is a much more difficult but not hopeless task. The goal of the binational agreement, though broadly stated, also includes specific objectives such as maintaining the lakes as a source of safe and high quality drinking water, as a safe place to swim, and as a source of high quality fish. It includes the maintenance and restoration of a lake environment that is suitable for native fish to reproduce and grow. It includes preservation of wetlands, prevention of excess sedimentation on spawning reefs, and prevention of structural development or power plant outfalls in the vicinity of spawning grounds.

The 1978 Agreement also contains specific objectives dealing with the control of toxic substances, excess nutrients, especially phosphorous, and the prevention of pollution arising from dredging and shipping actitivies.[29] Progress toward many of these goals has already been made. Complete achievement of the goals lies in the future, but at least agreement has been reached on the direction and purposes of management. The fact that a formal agreement between the U.S. and Canada has been reached is in itself a significant achievement. Its chances for success probably rest as much on a perception and recognition of mutual benefit as on any moral grounds or environmental ethic, though it can be argued that ecologically these are one and the same thing.

In contrast, the Law of the Sea negotiations are still bogged down by the attempts to combine the goals of preserving the marine environment and safeguarding its resources, with the goals of allocating immensely valuable mineral and biological resources, safeguarding the security of states, guaranteeing rights of passage, and preserving the freedom to conduct scientific research everywhere on the world ocean. The concept of the sea as a "commons" is simply not consistent with the present view of the resources of the sea as a source of profit for private industry and individual nations. That the integrity and preservation of the marine environment is essential for the long-term maintenance of the marine resource base has not yet been accepted by the majority of the 150 states involved in the LOS negotiations.

As in the Great Lakes Agreement, success of the LOS negotiations rests ultimately on the premise that there must be benefit to every group. The present prospect that such an arrangement can be reached in the foreseeable future appears to be dim indeed.

2. Great Lakes Fishery Commission

One of the most important and potentially effective Great Lakes management institutions is the Great Lakes Fishery Commission.[18] Established in 1955, the Commission is composed of the eight Great Lakes states, the Province of Ontario, and the federal governments of the U.S. and Canada.

Historically and legally the fishery resources of the Great Lakes are owned and managed by the individual states and Canadian provinces that border the lakes. As in any aquatic system neither the water nor the fish are influenced by state or national boundaries. Therefore what goes on in one jurisdiction can and does affect the fishery in others. The situation is aptly described by Fetterolf: "Management of the Great Lakes and the fishery resource is obviously complicated by the fact that the lakes are shared between Canada and the United States. What is less obvious is that the fishery management and regulatory authority rests with the eight states whose borders include parts of the lakes, and with the Province of Ontario, whose border includes parts of four of the lakes. This situation is most complex on Lake Erie where the international boundary line runs the entire length of the lake, some 400 km. The Province of Ontario is responsible for managing and regulating the fishery in the northern half. Starting from the west, the states of Michigan, Ohio, Pennsylvania, and New York manage and regulate the fisheries in each of their waters in the southern half.

This complexity and related situations on the other lakes have resulted in a series of efforts to develop effective, cooperative, management mechanisms."[18]

From 1893 to 1952, several unsuccessful attempts were made to create a joint management mechanism. In 1955, owing largely to the crisis of the sea lamprey ravages on the lake trout of the upper lakes, the effort finally succeeded.

In 1981 the members of the Commission reached major agreement on "A Joint Strategic Plan for Management of Great Lakes Fisheries."[31] The core of this plan is the goal statement which reads: "To secure fish communities, based on foundations of stable self-sustaining stocks, supplemented by judicious plantings of hatchery-reared fish, and provide from these communities an optimum contribution of fish, fishing opportunities, and associated benefits to meet needs identified by society for: wholesome food, recreation, employment, income, and a healthy human environment."[31]

The plan places heavy emphasis on the ecosystem base on which the fisheries depend. In order to protect fish habitats and ensure adequate spawning success the plan calls for aquatic environmental objectives the same as, or more stringent than, those of the environmental agencies. These objectives include protection of wetlands, control of eutrophication from excess nutrients, and the amelioration of chemical contamination by man-made organic chemicals like PCBs and mirex, and heavy metals like mercury, cadmium, and others.

The effort to manage the Great Lakes fisheries is basically one of ecosystem enhancement and the rehabilitation of threatened native species like the lake trout. Stocking of salmon and trout will continue, both as a means of controlling the alewife and to support the immensely valuable recreational fishery. Conflict between the commercial and recreational fishery interest and conflict over the claims of native Americans for a portion of the fishery exist and no doubt will continue. But the resolution of these conflicts is much further along than in the case of the coastal marine fisheries where such conflicts are just now arising.

B. State and National

State and provincial boundary lines join together in the middle of the lakes; there is no federal territory in the Great Lakes. The fish and the lake bed, including the minerals on or under the lake bed, are owned and controlled by the states. Who owns the water is much less clear. Regulatory authority to approve or disapprove diversions of water which affect the water level of the lakes is vested in the International Joint Commission. Small diversions have taken place on both sides of the international border without the approval of the IJC. Larger diversions have been vaguely discussed from time to time, but the authority of the IJC has never been truly tested.

As water needs grow outside the Great Lakes basin, more substantial diversions of Great Lakes water will certainly be proposed. How the states and federal governments will respond to these pressures remains to be seen.[27]

Water quality standards have been agreed upon under the binational Water Quality Agreement of 1978. These standards plus certain additional national standards and effluent regulations are enforced by the U.S. Environmental Protection Agency and Environment Canada. As in U.S. ocean coastal waters, navigation and dredging are entirely under federal control through the U.S. Coast Guard and the U.S. Army Corps of Engineers, respectively.

Authority for fishery management is vested entirely in the states and the Province of Ontario. Stocking, fishing regulations, and overall management policies including the balance between sport and commercial fisheries are state and provincial responsibilities. The Great Lakes Fishery Commission strives to coordinate these activities but it does not have regulatory or direct management authority except in the case of sea lamprey control.

Coastal management is a state responsibility. However, in the U.S. the Coastal Zone Management Act of 1972 provides a mechanism for a state and federal partnership in this function with the ultimate authority vested in the states and local jurisdictions. For a detailed discussion of this program see Chapter 7 by Miller and Chapter 8 by Owens.

In summary, management of the Great Lakes and their resources are shared between two nations, eight states and the Province of Ontario, and local jurisdictions. Though there are probably ambiguities remaining with regard to final authority over certain aspects of management, e.g., water diversion, there are no major gaps in the management structure.

V. THE LESSON OF THE GREAT LAKES

As a microcosm of the world ocean, the Great Lakes constitute a scientific, resource and management model of the ocean from which much can be learned. The ecosystem approach to the scientific study of the Great Lakes could well be adapted to the investigation of the world ocean. Global pollution, major ocean fisheries, particularly the migratory species, whales and other marine mammals, as well as ocean circulation, temperatures, ice cover, and climate interaction are all linked together by the ecosystem concept. Effective management of the world ocean must ultimately be based on a recognition of this concept.

Marine resources have long been exploited but are only now being recognized as finite and in need of management. This stage has already been reached in the Great Lakes and the development of management strategies is much further along than in the oceans. Environmental degradation, only now being experienced in the world ocean, has already occurred in some of the Great Lakes. The realization of the social and economic costs of such degradation of the Great Lakes has been recognized, and public support has enabled the government and its management agencies to undertake effective corrective measures. Such a realization has not yet generally occurred in most of the maritime nations of the world and as a result, little is being done to stem the process of continued degradation of oceans.

Finally the development of an institutional structure for the management of the Great Lakes is far more advanced than as is the case for the oceans. The obstacles to such an institutional structure are enormous as demonstrated by the slow and at times almost futile Law of the Sea negotiations. Even though these national and cultural obstacles cannot be eliminated by simple analogy, nevertheless much could be learned from the Great Lakes experience.

Man's relation to the marine environment has evolved from discovery and exploration to exploitation and degradation. Facing this latter phase, degradation, man has begun to realize the importance of understanding the marine environment. The threat of degradation and ultimate destruction of this part of our world can only be averted through an understanding of how it works, an understanding that goes beyond intellectual curiosity. The realization that the marine environment, and the Great Lakes portion of that environment, is an ecosystem and that man is a part of that ecosystem is just beginning to occur. The acceptance of this socially and politically revolutionary concept will not be easy nor will it occur naturally. Even in the Great Lakes *the idea* is struggling for acceptance. Its apparent abstractness strikes managers as idealistic and therefore not useful as a basis for planning. We will probably ultimately have to revise our institutional arrangements before the ecosystem concept can be effectively implemented. Even though it remains incomplete, the Great Lakes experience has been instructive. The repeated failures of the single resource approach to management has alerted the public to the wholeness of the system. We can only hope that our political leaders and resource managers will, in time, also learn this lesson.

REFERENCES

1. **Ragotzkie, R. A.,** The Great Lakes rediscovered, *Am. Sci.*, 62(4), 454, 1974.
2. **Ragotzkie, R. A.,** The Keweenaw Current, A regular feature of summer circulation of Lake Superior, Tech. Rep. 29, Nonr 1202(07), Department of Meteorology, University of Wisconsin, 1966, 30.
3. **Smith, N. and Ragotzkie, R. A.,** A comparison of computed and measured currents in Lake Superior, in *Proc. 13th Conf. Great Lakes Res.*, International Association Great Lakes Research, Ann Arbor, Mich., 1970, 969.
4. **Yeske, L. A.,** The structure of the Keweenaw Current by Photogrammetry, Ph.D. thesis, University of Wisconsin-Madison, 1973, 399.
5. **Green, III and Yeske, L. A.,** Horizontal Turbulent energy transfer associated with a Great Lakes coastal current, *Tellus*, 27(4), 384, 1975.
6. **Niebauer, H. J.,** Wind Driven Coastal Upwelling in Lake Superior, Ph.D. thesis, University of Wisconsin-Madison, 1976, 264.
7. **Niebauer, H. J., Green, T. III, and Ragotzkie, R. A.,** Coastal upwelling/down welling cycles in southern Lake Superior, *J. Phys. Oceanogr.*, 7(6), 918, 1977.
8. **Ragotzkie, R. A.,** personal observation, 1966—82.
9. **Stommel, H.,** *The Gulf Stream: A Physical and Dynamical Description*, 2nd ed., University of California Press, Berkeley, Calif., 1965, 248.
10. **Harrington, M.,** Currents on the Great Lakes, U.S. Weather Bureau, Bulletin B, Washington, D.C., 1895.
11. **Ragotzkie, R. A. and Bratnick, M.,** Infrared temperature patterns on Lake Superior and inferred vertical motions, in *Proc. 10th Conf. Great Lakes Res.*, University of Michigan, Great Lakes Research Division, Publication 13, Ann Arbor, Mich., 1966, 349.
12. **Ragotzkie, R. A. and Niebauer, H. J.,** Temperature regime in Lake Superior at Silver Bay, Minnesota, *Verh. Internat. Verein Limnol.*, 19, 144, 1975.
13. **Ragotzkie, R. A.,** Vertical motions along the north shore of Lake Superior, in *Proc. 17th Conf. Great Lakes Res.*, International Association Great Lakes Research, 1974, 456.
14. **Mortimer, C. H., (F.R.S.) and Fee, E. J.,** Free surface oscillations and tides of Lake Michigan and Superior, *Phil. Trans. Royal Soc. London*, A281, 1, 1976.
15. **Rao, D. B., Mortimer, C. H., and Schwab, D. L.,** Surface normal modes of Lake Michigan: calculations compared with spectra of observed water level fluctuations, *J. Phys. Oceanogr.*, 6(4), 575, 1976.
16. André, Father Louis, Extrait d'une lettre du Pere André escritte de la Baye des Puants, le 20 Avril 1676, in *The Jesuit Relations and Allied Documents*, Vol. 60, Thwaites, R. G., Ed., Burrows Bros., Cleveland, Ohio, 1900.
17. **Applegate, V. C.,** The sea lamprey, *Sci. Am.*, 4, 36, 1955.
18. **Fetterolf, C. M., Jr.,** Why a Great Lakes fishery commission and why a sea lamprey symposium, *Can. J. Fisheries Aquatic Sci.*, 37(11), 1588, 1980.
19. Sea Lamprey Control Program, Great Lakes Fishery Commission, U.S. Fish and Wildlife Service, Department of Fisheries and Oceans, Canada, U.S. Department of Interior pamphlet, 1981.
20. **Smith, B. R.,** Sea lampreys in the Great Lakes of North America, *The Biology of Lampreys*, Vol. 1, Hardisty, M. W. and Potter, I. C. Ed., Academic Press, N.Y., 1971.
21. **Delfino, J. J.,** Toxic substances in the Great Lakes, *Environ. Sci. Technol.*, 13(12), 1462, 1979.
22. **Talhelm, D. R., Bishop, R. C., Cox, K. W., Smith, N. W., Steinnes, D. N., and Tuomi, A. L. W.,** Current estimates of Great Lakes fisheries values: 1979 Status Report, Great Lakes Fisheries Commission, Ann Arbor, Mich. 1979, 18.
23. **Beeton, A. M.,** Eutrophication of the St. Lawrence Great Lakes, *Limnol. Oceanogr.*, 10, 240, 1965.
24. Water Quality Board, 1980 Report on Great Lakes Water Quality, International Joint Commission, Washington, D.C., 1980, 68.
25. Great Lakes Basin Commission, An Overview of Water Quality Trends and Issues in the Great Lakes States, Ann Arbor, Mich., Nov. 1980.
26. Eisenreich, S. J., Looney, B. B., and Thornton, J. D., Assessment of Organic Contaminants in the Great Lakes Ecosystem, Appendix, Annual Report, Science Advisory Board to International Joint Commission, Washington, D.C., 1980, 150.
27. State of Wisconsin, Coastal Management Program, *Proc. Conf. Interbasin Transfer of Water: The Great Lakes Connection*, May 1982, Milwaukee, Wisconsin, 1982, 221.
28. International Joint Commission, 70 Years of Accomplishment, Report for years 1978—79, Washington, D.C., 1979, 64.
29. Great Lakes Water Quality Agreement of 1978, International Joint Commission, Canada and the U.S., Nov. 22, 1978.
30. Great Lakes Advisory Board, The Ecosystem Approach, International Joint Commission, Washington, D.C., 1978, 47.
31. Great Lakes Fishery Commission, A Joint Strategic Plan for Management of Great Lakes Fisheries, Ann Arbor, Mich., 1980, 13.

INDEX

D

E

F

G

H

I

J

K

L

M

N

O

P

R

S

T

U

V

W

Y

Z